Forensic Science Laboratory Manual and Workbook

Third Edition

Thomas Kubic
Nicholas Petraco

 **CRC Press**
Taylor & Francis Group
Boca Raton London New York

CRC Press is an imprint of the
Taylor & Francis Group, an **informa** business

CRC Press
Taylor & Francis Group
6000 Broken Sound Parkway NW, Suite 300
Boca Raton, FL 33487-2742

© 2009 by Taylor & Francis Group, LLC
CRC Press is an imprint of Taylor & Francis Group, an Informa business

No claim to original U.S. Government works
Printed in the United States of America on acid-free paper
10 9 8 7 6 5

International Standard Book Number-13: 978-1-4200-8719-2 (Softcover)

Library of Congress Cataloging-in-Publication Data

Kubic, Thomas.
   Forensic science laboratory manual and workbook / Thomas Kubic and Nicholas Petraco. -- 3rd ed.
      p. cm.
   Includes bibliographical references.
   ISBN 978-1-4200-8719-2 (alk. paper)
   1. Forensic sciences--Handbooks, manuals, etc. 2. Criminal investigation--Handbooks, manuals, etc. 3. Evidence, Criminal--Handbooks, manuals, etc. 4. Crime laboratories--Handbooks, manuals, etc. I. Petraco, Nicholas. II. Title.

HV8073.K75 2009
363.25--dc22   2008048757

Visit the Taylor & Francis Web site at
http://www.taylorandfrancis.com

and the CRC Press Web site at
http://www.crcpress.com

Contents

Preface..xiii
The Authors xv
List of Equipment and Supplies xvii
General Forensic Laboratory Safety Rules xxi

Part 1

Experiment 1 Introduction to Scientific Measurement and Experimental
Error — Determining the Density of Glass 3

Teaching Goals...3
Background Knowledge..3
Equipment and Supplies 4
Hands-on Exercise 5
    A. Basic Measurements 5
Optional Exercise: The Variability of Measurements and the Standard Deviation.........5
    B. Further Observations on the Variation of Measurements: Weighing, Volume,
        and Density Determination 6
Optional Exercise: Determining the Density of a Glass Sample 7
Data Form 8
    A. Document Measurements and Observations...8
    B. Document Experimental Data...9
    C. Glass Density Determination 10

Experiment 2 Understanding Elements of Identification and Individualization 11

Teaching Goals...11
Background Knowledge..11
Equipment and Supplies 11
Hands-on Exercise 12
Report...15

Experiment 3 Use of the Compound Microscope 17

Teaching Goals 17
Background Knowledge 17
Equipment and Supplies 18
Hands-on Exercise 18
Alignment Instructions 18
Köhler Illumination 19
Report 21

Experiment 4 Stereomicroscopes and Firing Pin Impressions (Tool Marks) 23

Teaching Goals 23
Background Knowledge 23
Equipment and Supplies 24
Hands-on Exercise 24
    Part I: The Stereomicroscope 24
    Optional Exercise 25
    Part II: Observations with the Stereomicroscope 26
    Part III: Comparison of Firing Pin Impressions 26
Report 29

Experiment 5 Acquiring and Classifying Inked and Latent Fingerprints 31

Teaching Goals 31
Background Knowledge 31
Equipment and Supplies 32
Hands-on Exercise 33
Report 37
Appendix A 39
Appendix B 40

Experiment 6 Identification and Matching of Fingerprints 41

Teaching Goals 41
Background Knowledge 41
Equipment and Supplies 41
Hands-on Exercise 42
Some Important Features and Their Definitions 42
Report 45

Experiment 7 Cyanoacrylate "Super Glue" Fuming Technique to Develop Latent
               Fingerprints 47

Teaching Goals 47
Background Knowledge 47
Equipment and Supplies 47
Hands-on Exercise 48
Report 51

Experiment 8   Crime Scene Investigation:
Safeguarding, Searching, Recognition, Documentation,
Collection, Packaging, and Preservation of Physical Evidence
Found at the Crime Scene 53

Teaching Goals 53
Background Knowledge 53
Equipment and Supplies 53
Hands-on Exercise 54
Important Concepts and Observations 55
Crime Scene Activities 55
Report 59

Experiment 9   Trace Evidence Collection and Sorting 61

Teaching Goals 61
Background Knowledge 61
Equipment and Supplies 61
Hands-on Exercise 62
Report 65
Appendix A 67

Experiment 10   Sample Preparation for Microscopic Examination 69

Teaching Goals 69
Background Knowledge 69
Equipment 69
Hands-on Exercise 69
      Preparation of Wet Mount in Permount® 70
      Preparation of Scale Cast Impression in Meltmount® 70
      Preparation of Wet Mount in Meltmount 71
      Preparation of Hair or Fiber Cross Sections on Plastic Slides 72
Report 73

Experiment 11   Examination of Human Hair 77

Teaching Goals 77
Background Knowledge 77
Equipment and Supplies 77
Hands-on Exercise 78
Report 83
Appendix A: Human Hair Data Sheet 85
Appendix B: Glossary of Hair Terms 85

Experiment 12   Examination of Mammalian Hair 87

Teaching Goals 87
Background Knowledge 87
Equipment and Supplies 87
Hands-on Exercise 87

Report..91
Appendix A 93
Appendix B 93
Appendix C 94

Experiment 13 Measurement with the Microscope 95

Teaching Goals..95
Background Knowledge...95
Equipment and Supplies 95
Hands-on Exercise 96
Report..99

Experiment 14 Examination of Trace Quantities of Synthetic Fibers 101

Teaching Goals.. 101
Background Knowledge... 101
Equipment and Supplies 101
Hands-on Exercise 101
Report.. 107
Appendix A 109

Experiment 15 Basics of Photography 111

Teaching Goals.. 111
Background Knowledge... 111
Equipment and Supplies 111
Hands-on Exercise 111
Report.. 115
Photography Glossary 117

Experiment 16 Black-and-White Film Development 119

Teaching Goals.. 119
Equipment and Supplies 119
Hands-on Exercise 119
Film Development.. 119
Enlargement of Negatives 120
Print Development 121
Report.. 123

Experiment 17 Collection of Footwear Evidence 125

Teaching Goals.. 125
Background Knowledge... 125
Equipment and Supplies 125
Hands-on Exercise 126
Report.. 131

Experiment 18 Identification and Comparison of Footwear Impressions 133

Teaching Goals...133
Background Knowledge...133
Hands-on Exercise 133
    Preparation of Known Standards..133
    Questioned Print Examination...134
Important Concepts..134
Evaluation Process...135
Conclusions..136
Report..137
Footwear Glossary 140

Experiment 19 Tool Mark Examination...141

Teaching Goals...141
Background Knowledge...141
Equipment and Supplies 141
Hands-on Exercise 141
    Questioned Tool Mark Examination..141
    Preliminary Examination of Questioned Tool Marks..142
    Preparation of Known Standards..142
    Examination and Comparison of Questioned and Known Tool Marks......................145
    Evaluation Process...146
Conclusions..146
Report..147

Experiment 20 Glass Fractures and Direction of Force 149

Teaching Goals...149
Background Knowledge...149
Equipment and Supplies 149
Hands-on Exercise 150
Report..151

Experiment 21 Thin-Layer Chromatography:
                         Separation of Dyes in Ballpoint Inks...155

Teaching Goals...155
Background Knowledge...155
Equipment and Supplies 155
Hands-on Exercise 156
Report..157

Experiment 22 Bloodstain Geometry (Part A)...159

Teaching Goals...159
Background Knowledge...159
Equipment and Supplies 159
Hands-on Exercise 160
Artificial Blood Preparation..160

Report.. 161
Bloodstain Pattern Glossary.. 163

Experiment 23   Bloodstain Geometry (Part B) 165

Teaching Goals... 165
Equipment and Supplies 165
Hands-on Exercise 165
Report.. 169

Experiment 24   Forgery Detection 171

Teaching Goals... 171
Background Knowledge.. 171
Equipment and Supplies 171
Hands-on Exercise 171
Report.. 175
Appendix A: Important Features to Examine 177

Experiment 25   Soil Examination 179

Teaching Goals... 179
Background Knowledge.. 179
Equipment and Supplies 179
Hands-on Exercise 179
Report.. 183

Experiment 26   Forensic Odontology 1:
                    Is It a Bite Mark? 187

Teaching Goals... 187
Background Knowledge.. 187
Equipment and Supplies 187
Hands-on Exercise 188
Report.. 197
Glossary of Terms for Human Bite Marks.. 199

Experiment 27   Forensic Odontology 2:
                    Who Made the Questioned Bite Mark? 201

Teaching Goals... 201
Background Knowledge.. 201
Equipment and Supplies 201
Hands-on Exercise 201
    Examination of Knowns 203
    Examination of Questioned Marks 206
Report.. 211

Experiment 28   Forensic Archeology:
                    Search for Human Habitation and Remains 213

Teaching Goals... 213
Background Knowledge.. 213

Equipment and Supplies 213
Hands-on Exercise 214
Report..221

Experiment 29 Forensic Anthropology 1:
Basic Human Osteology 223

Teaching Goals..223
Background Knowledge..223
Equipment and Supplies 223
Hands-on Exercise 223
Glossary 240

Experiment 30 Forensic Anthropology 2:
Examination of Grave Site Bones..243

Teaching Goals..243
Background Knowledge..243
Equipment and Supplies 243
Hands-on Exercise 243
Report..253

Experiment 31 Digital Photography Image Processing 255

Teaching Goals..255
Background Knowledge..255
Equipment and Supplies 255
Hands-on Exercise 255

Part 2

Experiment 32 Chromatography 2:
Identification of a Single-Component Solvent by Gas Chromatography ...265

Teaching Goals..265
Background Knowledge..265
Equipment and Supplies 271
Hands-on Exercise 271
Optional Exercise..272
Report..273

Experiment 33 Spectroscopy 1:
Methods for the Identification of Materials Based
on Their Absorption of Light of Various Wavelengths
and Identification of Polymer Films by FTIR..275

Teaching Goals..275
Background Knowledge..275

Equipment and Supplies 279
Hands-on Exercise 279
Report.. 281

Experiment 34   Spectroscopy 2:
                    Use of Visible Spectroscopy in Color Determination 283

Teaching Goals.. 283
Background Knowledge... 283
Equipment and Supplies 283
Hands-on Exercise 284
Optional Exercise.. 284
Report.. 285

Experiment 35   Quantitative Ethyl Alcohol Determination by Gas Chromatography 287

Teaching Goals.. 287
Background Knowledge... 287
Equipment and Supplies 288
Hands-on Exercise 288
Optional Exercise.. 288
Report.. 291

Experiment 36   Forensic Entomology.. 295

Teaching Goals.. 295
Background Knowledge... 295
Equipment and Supplies 297
Procedure 298
Report.. 301
Appendix A 303
Appendix B 304
References.. 305

Experiment 37   Basic Ballistics 307

Teaching Goals.. 307
Background Knowledge... 307
Breech Marks 309
Equipment and Supplies 309
Procedure 309
Report.. 317
References.. 319

Experiment 38   Basic Ballistics 2 321

Teaching Goals.. 321
Background Knowledge... 321
Report.. 333

Experiment 39 Crime Scene Drawing with Microsoft Word® 335

Teaching Goals..335
Background Knowledge..335
Materials 335
Procedure 335
    Toolbar Option Selections 336
    File: Page Setup Options 336
    View: Toolbar Setup Options...336
    Drawing Toolbar Settings and Options...336
    Steps in Preparing a Crime Scene Drawing..337
Reference 337
Appendix A 342

Appendix 1   Reference Data for Polymer Films 345

Bibliography 359

Preface

This workbook/laboratory manual is designed for a cooperative learning setting in which three to five students comprise a group. Although every student is required to maintain his or her own laboratory notebook, each student in the group is responsible for different tasks in a given exercise, while the group as a whole works together to complete the assignment. It is suggested that the tasks for each exercise be assigned by a method of random chance so that each student will develop skills in a number of the procedures, abilities, and processes covered in this text.

Hands-on learning is an integral part of the exercises in this workbook. Tasks such as measuring the positions of objects using polar coordinates or triangulation, taking photographs and videos, sketching scenes, using a compass, drawing to-scale diagrams, and using sophisticated scientific equipment to collect and examine various types of evidence are mandatory. In addition, using and understanding the scientific methods of reasoning, deduction, and problem solving are essential in order to carry out all of the experiments and exercises covered.

The Authors

Thomas Kubic, JD, Ph.D., FABC is currently a professor of forensic chemical instrumentation, scientific and expert testimony, electron microscopy, and advanced trace evidence analysis at John Jay College of Criminal Justice, City University of New York (CUNY), and director of TAKA Instructional Agency, a New York State not-for-profit educational institution concentrating on training in microscopy. He holds a Ph.D. with a concentration in forensic science from CUNY, a master's degree in chemistry from Long Island University, and a law degree from St. John's University. He has been admitted to the New York State Bar. Dr. Kubic has been awarded the Paul L. Kirk Award (1997) for contributions to criminalistics and forensic science by the Criminalistics Section of the American Academy of Forensic Sciences and the Arthur Neiderhoffer Scholarship for significant contributions to the field of criminal justice by the Graduate School City University of New York. He retired from a municipal crime laboratory where he spent over 20 years as a forensic microscopist, is a member of the FBI-sponsored SWGMAT committee on forensic glass analysis, is a technical expert with NIST's laboratory accreditation program (NVLAP), and has been the laboratory director of a New York State accredited environmental laboratory. Dr. Kubic has been an instructor in a number of microscopy and industrial hygiene courses at the continuing education program of EOSHI, a division of UMDNJ, Rutgers University. His research interests are in the application of light and electron microscopy with digital image techniques to the analysis of particulates and microscopic forensic evidence, as well as mentoring graduate students in criminalistics research at John Jay College. He maintains a successful consulting company dealing with the identification and characterization of particulate materials by light and electron microscopy, especially when of forensic import. Dr. Kubic has qualified as an expert and testified to his opinions in both federal and state courts in criminal and civil actions. He is the author of 20 scientific and technical articles as well as a number of textbook chapters, and has made over 60 presentations at technical conferences. He is the coauthor of an atlas on the application of light microscopy for criminalists, chemists, and conservators and a laboratory manual for beginning forensic science students. Dr. Kubic is a Fellow of the American Academy of Forensic Sciences and holds Fellow status as a certified criminalist. He is a member of the New York Microscopical Society, the Microscopy Society of America, the Microbeam Analysis Society, the American Chemical Society, and other professional organizations.

Nicholas Petraco earned a B.S. in chemistry and an M.S. in forensic science from John Jay College of Criminal Justice, the City University of New York. He served as a detective/criminalist at New York City's Police Laboratory from 1968 to 1990 and held the position of senior forensic microscopist of the laboratory's trace section between 1982 and 1990, when he became a private forensic consultant. In addition to participating in the professional activities listed, Mr. Petraco has been the technical leader for the trace evidence and criminalistics sections of the NYPD forensic investigation division since 1999. His duties include doing casework, as well as advising, teaching, and lecturing members of both the forensic laboratory and crime scene unit.

Throughout his career, he also served as a lecturer at John Jay College and was an associate professor at St. John's University. Currently, he is a full-time assistant professor at John Jay College with the science department. Mr. Petraco has helped educate thousands of forensic scientists, worked on more than 5000 death investigations on behalf of prosecution and defense attorneys, and testified as an expert in more than 500 trials conducted in local, state, and federal criminal and civil courts.

Mr. Petraco is a Fellow of the New York Microscopical Society (FNYMS), a Fellow of the American Academy of Forensic Scientists (FAAFS), and a Diplomate of the American Board of Criminalistics (DABC). He served as chairperson of the SWGMAT forensic hair committee from 1997 to 2001. He has presented over 100 papers at scientific meetings and symposia, has published over 50 articles in the forensic literature, and has authored and coauthored six book chapters and three textbooks and CDs on various subjects in forensic science, microscopy, and crime scene examination.

List of Equipment and Supplies

An inventory of the materials needed for the forensic laboratory experiments outlined in this manual is provided below. Each item listed should be made available to each group of participating students, except as otherwise noted.

- Bright field compound microscopes with 4×, 10×, 20×, and 40× achromatic objective lenses, 10× oculars (eyepieces), and an in-base illumination system with focusable condensers
- Two polarizing microscopes, one with a trinocular head (shared by students)
- Stereomicroscopes with a magnification range of 4× to 40×
- Eyepieces with reticle scales for stereomicroscopes and compound microscopes, or have reticles installed in existing units
- Stage micrometers for reflected light microscopes or high-quality scale in millimeters (can be shared by students)
- Stage micrometers for compound microscopes, 1 mm in 1/100 divisions (can be shared by students)
- Hand-held magnifying glasses, 5× to 10×, or linen testers
- One-half gross of 25 × 75 mm precleaned glass microscope slides
- Assortment of no. 1 1/2 glass cover glasses: square 18 mm, rectangular 22 × 40 mm and 22 × 50 mm, round 18 mm (shared)
- Plastic microscope slides
- Assortment of mounting media: distilled water, Permount, Meltmount 1.539, Cargille's refractive index oil set A (mounting media can be shared by all students)
- Plastic rulers, protractors, and tape measures 1 to 50 ft metric/English
- Two fine, nonmagnetic, stainless steel tweezers
- One trace evidence vacuum with 25 traps (shared)
- Two rolls of dust lifting tape
- Two stainless steel dissecting needles with holders
- No. 11 scalpel blades
- Microspatula
- One box of flat wooden toothpicks
- Small scissors with pointed tip
- Equipment for thin-layer chromatography: 25 × 75 mm and 5 × 25 cm silica gel TLC plates, reagents (sprays, solvents), glassware, TLC tanks with covers, 12 Coplin jars with covers, blotter paper, microspotting pipettes
- Ultraviolet light, both short and long wavelength (hand held, these are good for searching for evidence)
- One platinum loop with holder
- Small magnets
- pH test paper, pH 0 to 14
- One thermometer, temperature 0 to 100°C
- One alcohol lamp

- Balances, 1 mg to 1000 g (can be shared by students)
- Assorted micro and macro disposable pipettes
- Aluminum trays or cardboard boxes (width 12 to 24 in. × length 24 to 36 in. × depth 6 in.)
- Aluminum weighing dishes, 60 mm diameter (100 per semester)
- Glass graduated cylinders 1 to 50 mL, subdivisions 0.1 to 1 mL depending on its cylinder
- Disposable petri dishes, 75 to 125 mm
- Cellophane tape
- Plastic bones
- Plastic skulls with teeth
- Sieves
- Computer
- Adobe Photoshop software (Adobe Elements 3.0 will suffice)

Photography Equipment

- Three 35 mm cameras with flash (detachable) and lenses
- Digital camera, at least 3.0 megapixels
- Computer, color printer, ink jet photographic paper, 64 to 128 MB storage card and digital storage card reader
- Photographic scales, tripods, black-and-white film, color film (process commercially), polaroid film if needed, and miscellaneous camera accessories
- Darkroom and equipment for black-and-white film development and printing, including, but not limited to, an enlarger, photographic paper, chemicals, and necessary equipment. (This assumes you have a room with electric and water that can be turned into a darkroom.)
- If no darkroom, then two-speed or crown Graflex cameras (used) with 4 × 5 Polaroid backs

Fingerprint Supplies (shared)

- Fingerprint feather dusters (12)
- Fiberglass latent brushes (12)
- Magnetic applicator brushes (4)
- Assortment of dusting powders: white, black, magnet, and fluorescent
- Lifting tape with dispensers, 2 in. (12 rolls)
- Black-and-white gloss card pads, 50 cards per pad: 3.5 × 5 in. and 8.5 × 11 in. (4 each)
- 2 × 2 in. hinged lifters, white (100)
- Super Glue fuming material for fingerprints: glue, warming plates, glue dishes
- Fingerprint ink pads (5)
- Fingerprint cards (250)
- Fingerprint stands, with pads and rollers (2)
- An alternative forensic light source
- An ultraviolet light source

Footprint Processing Equipment (shared)

- Electrostatic dust print lifter (1)
- Dental casting material, 25 lb box
- Plastic bottles for distilled water, 1 pt. (6)
- Clear footwear lifting sheets with covers (250)
- Any old footwear

Tool Mark Supplies (shared)

- Assorted hand tools: wire cutters, cable cutters, saws, drills, screwdrivers, leatherman tool, and so on
- Assorted locks
- Pieces of wood and metal
- Aluminum rods and plates
- Silicone casting material

Evidence Packing Equipment for Teaching Chain of Custody

- Evidence envelopes
- Evidence boxes
- Evidence bags
- Evidence containers
- Evidence tape

General Forensic Laboratory Safety Rules

Objective: To make available a safe and academically beneficial environment in the forensic laboratory by promoting safety and alertness.

- You must wear approved safety eyeglasses in the laboratory at all times. This means eye protection that will protect against both accidental splashes of chemicals, accidental explosions, and blunt impacts. Normal eyeglasses do not meet these requirements.
- Wear protective clothing in the laboratory.
- You must notify the instructor immediately of any accident, fire, or dangerous condition.
- You must know the location of and know how to operate all laboratory safety equipment, including eyewash, safety shower, fire hose, fire extinguisher, fire blanket, safety shower, and neutralizing chemical agents such as potash.
- If a chemical splashes into your eyes, you must immediately wash your eyes with cold water for at least 30 min. Use freely flowing cold water from the safety eyewash or sink.
- Immediately extinguish any fires to person or property. Remove all affected clothing while drenching the fire under the safety shower. Use wet towels or fire blankets to suffocate the fire.
- Immediately neutralize any chemical spills with neutralizing agents located in the hood or in spill buckets.
- When working with chemicals, always work under a ventilation hood to avoid inhaling potentially harmful fumes.
- Secure, tie up, or cover long hair when in the laboratory to prevent accidental burning of hair or getting hair caught in equipment.
- Before use, wash all laboratory glassware. Never add any chemicals to soiled glassware.
- Keep the laboratory and your working area clean and free from clutter.
- Eating and drinking of foodstuffs is not allowed in the laboratory.
- No smoking is allowed in the laboratory.
- Protect against creating a biological hazard. Many biological materials are hazardous. Wear protective clothing, laboratory coat, apron, or other protective outer clothing that is removed immediately after leaving the laboratory. Wash your hands thoroughly after handling anything in the laboratory.
- You must never perform an unauthorized experiment or procedure. Do not deviate from the instructions given by your instructor or written in the laboratory manual. Never perform a laboratory procedure or experiment you are uncertain about (ALWAYS ASK THE INSTRUCTOR).
- Never pipette by mouth. Always use a suction bulb.
- Do not force glass tubing into rubber stoppers. Lubricate the tubing and protect your hands with puncture-resistant gloves.
- Never work in the laboratory alone.
- Immediately seek medical attention for cuts, burns, and other serious injuries.

Part 1

Introduction to Scientific Measurement and Experimental Error — Determining the Density of Glass

Teaching Goals

The student will become familiar with the collection of scientific data and the performance of measurements. The student will learn about the basis for the uncertainty in measured values and the types of error that can arise. He or she will make a number of measurements and determine values based on these measurements. Each student will perform a number of basic statistical calculations and compare the data obtained from a single experimenter and data obtained from a group.

Background Knowledge

The purpose of this experiment is to demonstrate the uncertainty of scientific measurements due to a number of possible errors in the measurement process. First, the student will make a number of measurements, including length, weight, and liquid volume determinations. Second, the student will determine the variation in measurements of the same items by multiple students. The student will, as an option, determine the standard deviation (variability) of these measurements. Third, as an additional optional exercise for this experiment, the student may determine the density of glass by measuring the mass and volume of glass shards.

The general public tends to believe that when a scientist makes a measurement, especially a very simple one, there are no errors. This is simply not the case. Rather, it is a fact that all scientific measurements of a quantitative nature are imperfect and have an uncertainty associated with them. The goal of scientists is to do their utmost to decrease this uncertainty as much as possible. This uncertainty is a combination of all the "errors" involved in making a measurement.

When we speak about errors, we seldom mean "mistakes" like transcribing the wrong value, arithmetic miscalculations, or spilling part of the contents of a vessel that is our test sample. Here we mean the errors that occur due to imperfections in our apparatus, its calibration, or random variation due to our inability to reproduce a particular portion of the procedure being used to make the measurement. The determination of the total uncertainty of a measurement is not necessarily a simple matter. As an introduction, we will present a number of basic concepts in this experiment with the goal of demonstrating them in the laboratory exercise.

There are three general classifications of error by scientists. Some have subdivisions. These are systematic, random, and gross errors. Gross errors, sometime called "blunders," are those errors that are caused by less than careful attention to the task at hand. All the items we mentioned above as mistakes, as well as inattention to reading a value from an instrument, misweighing a sample, and others all fall into this class. The cause of the error, in most cases, unless noticed and reported by the analyst, cannot be traced. The only way to ensure that a blunder has not taken place is to make several measurements or determinations. If a blunder takes place, the suspect value will appear very different from the others. There is a wealth of literature dealing with the identification of such "outlier" values and their rejection based on sound statistical treatment. Arithmetic miscalculations and data entry errors into spreadsheets are exceptions to this rule of multiple determinations, but some type of hard copy record must be available for examination to reveal the origin of these errors.

Systematic errors are those that are consistently repeated each time a particular instrument or device is employed to make a measurement. It is generally agreed that systematic error can be eliminated by proper calibration procedures. Systematic errors can be absolute or relative. Absolute error means that each time a measurement is made, a specific error occurs. For example if a tape measure is 1/4-in. short at the front of the tape, each measurement will be 1/4-in. short no matter if one is measuring 1 in. or 10 ft. Relative errors are normally expressed as percentages (%) and affect the value measured by the same relative amount no matter what the total value, that is, if one had a 1-L flask that really contained 980 mL, it would have a 2% relative error. If we used it to measure 10 L by refilling and emptying it, the relative error would remain at 2%.

Random error occurs when a measurement is made and specific actions cannot be performed perfectly because of human factors or equipment imperfections. These errors are not the same each time they occur. Reading the meniscus on a burette or pipette, wherein a slightly different amount is delivered each time it is used is an example of how random errors are generated.

The true value (in analysis) is the value that characterizes a quantity perfectly under the conditions that exist when that quantity is considered. It is an ideal value that can be arrived at only if all causes of measurement error are eliminated and the entire population is sampled when calculating the mean of all measurements. The true value is also considered to be one that is calculated from fundamental principles, determined by a number of unrelated measurement techniques that agree, or established by definition. The arithmetic mean of a measurement calculated from a reasonable number (at least seven) and properly picked samples from a full population is often considered the true value.

Equipment and Supplies

1. Plastic scale or rulers with numbered inches and millimeter scales
2. Plastic protractor
3. Triple beam balance, 0.01 g
4. Glass pipettes, 10 mL to delivery
5. Disposable aluminum or plastic dishes (60 × 15 mm)
6. Glass graduated cylinders, 25 mL or 50 mL
7. Deionized or distilled water
8. Assorted beakers
9. Pieces of glass from the sides and bottoms of broken bottles (1 × 1 cm square)
10. Tape measures: 12 ft divided into 1/16-in. increments; 25 ft. divided into 1/8-in. increments

Hands-on Exercise

A. Basic Measurements

Figure 1.1

Calculate the average of your and at least two classmates' data from the measurement data recorded on either the blackboard or the class data sheet using the following equation:

$$\bar{x} = x_i / n$$

where $\bar{x}$ = the average, x_i = the sum of all data points, and n = the total number of data points.

Employing the 12-ft. tape, measure an object designated by your instructor (e.g., a desk or a blackboard) to the nearest 1/16 in. and record the dimensions below. Employing a longer tape, measure a large area designated by the instructor. Record the dimensions. Calculate the average of the measurements made by all students in your class employing data from the Data Sheet, Table 1.2.

Optional Exercise:
The Variability of Measurements and the Standard Deviation

Determine the variability of one or more of the groups of the measurements in columns 1 to 9 by calculating the standard deviation of each measurement using the following equation:

$$S = \left[\sum (\bar{x} - x_i)_j^2 \Big/ n - 1 \right]^{1/2}$$

The instructor will explain how to perform the calculation manually and the meaning of standard deviation. Other equations are valid and may be used. You may use a scientific calculator or computer spreadsheet. Make, fill in, and use a table similar to the one immediately below if you are performing your calculations manually.

TABLE 1.1
Standard Deviation Calculation Data

Measurement Number	Measurement Value	$(\bar{x} - x_i)$	$(\bar{x} - x_i)^2$
1			
2			
3			
$n = 3$			$\Sigma(0-x_i)^2_j$

$(\bar{x} - x_i)$ = the difference between the average of the group and any particular value.

$(\bar{x} - x_i)^2$ = the square of the difference between the average and any particular value.

$\Sigma(\bar{x} - x_i)^2_j$ = the sum of all the squares of all the differences.

B. Further Observations on the Variation of Measurements: Weighing, Volume, and Density Determination

Mass = the quantity of material that makes up an item
Gram (g) = the unit of mass in the CGS (metric) system
Weight = the gravitational force acting upon a body at the Earth's surface*
Density (D) = mass per unit volume (g/mL)

A) Obtain an item (or items) from the instructor and weigh it (them) to the nearest division readable on your balance. Record the weights and a brief description of the items on Data Sheet 1. Calculate the average for each item based on your data and that of at least three classmates. Using all the data from the class would be better. Calculate the standard deviation as an option either manually, with a calculator, or with a spreadsheet.
B) Obtain a weighing boat and weigh it empty. (This is called taring the container.) Using a high-quality glass pipette, transfer two or three 10 mL portions of water into the boat (depending on the size of the boat) by filling, emptying, refilling, and emptying, etc., the pipette. Reweigh the boat with the water in it. Determine the weight of the water by subtracting the tare weight of the boat. Record your data in the lab book.

Assuming that water has a density of 1 g/mL, calculate the volume of water you transferred.

Density (D) = m(ass) in g(rams)/v(olume) in mL

Compare this with the values that at least three fellow students obtained. To remove a variable, compare only values obtained from the same number of pipettings. Calculate an average and standard deviation as previously. Calculate the percent (%) error of your determination and compare it to that when using the average value.

Percent error is equal to the absolute value (a value neglecting the arithmetic sign) of the difference between the true value (the accepted value from references) and the value determined in your experiment divided by the true value, with this quantity multiplied by 100 and reported as a percentage.

Percent error = absolute value of [(true value – determined value)/true value] * 100

---

* In common terminology when we "weigh" something we are really determining its mass.

For our experiment, the true value is the total volume transferred times the reference value for the density of water. This can be assumed to be one, but it should be corrected for temperature (from Table 1.2) if the teacher instructs you to do so.

Percent error = [(mL actually transferred (from the weight) – volume from pipetting)/

mL actually transferred (from the weight)] × 100

Optional Exercise: Determining the Density of a Glass Sample

Obtain, clean, and dry a few pieces of broken bottle glass. Soda or beer bottles work well and the broken samples should, for safety, be prepared by the teacher. Obtain a "To Contain" graduated cylinder if available, and fill to approximately half its volume with water and record the volume to the nearest 0.2 mL or better. See Figure 1.2 on the proper manner of reading the meniscus of a liquid in a cylinder or burette.

Weigh one or more pieces of glass totaling 3 to 6 g to the highest readability of your balance. Carefully place (don't splash) the glass pieces into the graduated cylinder and record the new volume. The change in the volume is the volume of water displaced by the glass and is the volume of the glass.

Determine the volume change ($V_{end} - V_{start}$), which is the volume of the glass (V_{glass}).

Calculate the density of the glass: $D_{glass} = m_{glass}/V_{glass}$.

D = mass (weight) of glass in grams/volume of water displaced

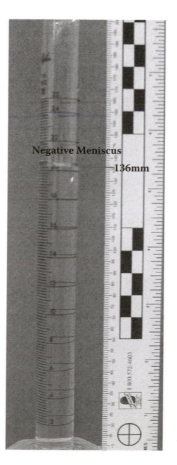

Negative Meniscus

136mm

gure 1.2

TABLE 1.2
Data Sheet

Enter measurements in the proper spaces. Note equipment identification in parentheses [e.g., for Side 1: 22.5 mm (B)]. Enter measurement/weight data on the lines next to your student number. Enter your classmates' data on the lines next to their numbers.

Student	Side 1	Side 2	A	B	Desk	Floor	Weights A	B	C	Eq. 1.3*	Eq. 1.4**
1											
2											
3											
4											
5											
6											
7											
8											
9											
10											

*Equation 1.3, weight in grams of pipetted water/mL of pipetted water.
**Equation 1.4, density of glass = mass in grams/volume in milliliters.

Data Form

Use the data sheet in Table 1.2 to record the data from your work and your classmates' as per the instructor's directions. Transfer the appropriate data to the report section that follows the data sheet.

Student Name _____ Date _____

Fill out the remainder of this form, a copy of which is to be submitted to your instructor for evaluation. Be sure to clearly identify the measurement and the value which you are reporting.

A. Document Measurements and Observations

1. Length and angle measurements

2. Group or class averages and standard deviations

3. Mass (weight) determinations
   a.   Weights

   b.   Averages and standard deviations for the group or class

B. Document Experimental Data

1. Water transfer variation
   a.   Volume that should have been transferred (volume of pipette × number of times filled and emptied)

   b.   Actual volume transferred

      i. Your determination

      ii. Group or class average

c. Percent error determination

1. Your data

2. Group or class average

C. Glass Density Determination

1. Density, your determination

2. Density, group or class average and standard deviation

3. On a separate sheet, write a short narrative revealing what you learned from these exercises.

3

Use of the Compound Microscope

Teaching Goals

Students will learn how to align a compound microscope and set up proper illumination. Students will also learn the proper use of a compound light microscope.

Background Knowledge

The light microscope is one of the most valuable instruments in the crime laboratory. The basic transmitted light microscope is supplemented by the stereomicroscope, polarized light microscope, phase contrast microscope, and visible and infrared microspectrophotometers. In some cases these instruments are employed as the final analytical tool and in others they are used for preliminary examinations or for sample preparation.

These microscopes are referred to as compound microscopes because the final magnification — the apparent size of the image — is computed by multiplying the objective (primary, first lens) power by the eyepiece (ocular) power. For example, a very common combination is a 10× objective employed with a 10∞ eyepiece for a total magnification of 100× (10 × 10 = 100).

The quality of the image depends on the quality of the lenses employed in the instrument and the proper alignment of the microscope. Lenses range from very inexpensive (uncorrected) to moderately priced simply corrected (Acromats) to very expensive highly corrected (Apochromats).

The following terms are employed to describe certain important concepts involved in the use of the microscope:

Depth of field — the thickness of the parts of the sample that appear to be in focus (clear and sharp) simultaneously. The greater the magnification of an object, the less the depth of field.

Field of view — the size (area or diameter of visible field) of the specimen viewable at once.

Numerical aperture (NA) — the equation that describes the relationship of the angle (AA) of light that the objective collects and the refractive index (η) of the medium between the specimen and the objective. It is important to the determination of resolving power. NA = η × sin AA/2. NA varies with magnification for similar-quality objectives.

Resolving power (RP) — the smallest details separated by a distance equal in size to the details that can be determined to be separate objects. It is related to the NA, the wavelength (λ) of the light employed, and a constant (k), equal to 0.61, related to the sensitivity of the human eye. The smaller the RP, the smaller the observable detail. $RP = \dfrac{.61\lambda}{NA}$.

Refractive index (η) — the ratio of the speed of light in a vacuum divided by the speed in the medium of interest. For visible light, it is always equal to or greater than one.

Working distance — the distance from the top of the specimen (when in focus) to the lower tip of the objective. Generally the greater the magnification used, the shorter the objective's working distance.

18

In this experiment you will put into practice procedures for the proper alignment of the light microscope that were described previously in the lecture or lab. In addition, you will conduct a few observations of selected materials at various magnifications.

Equipment and Supplies

1. Transmitted light compound microscope with illuminator
2. Microscope slides and cover slips
3. Sample manipulation tools (dissection needles [2], index cards, weighing paper, weighing boats, scissors [small], tweezers, scalpel[s], cellophane tape, small plastic or glass vials with caps or corks, unsharpened pencils [2] with erasers)
4. Tissues and lens paper
5. Clove oil (RI.543), oil of cedar (RI.505), α-monobromnaphthalene (RI.66), or commercial oils with similar refractive values
6. Set of prepared (previously mounted) reference materials of forensic interest for viewing
7. Human hair, tea, tobacco, synthetic fiber, cotton

Hands-on Exercise

1. Review your notes on microscope alignment.
2. Refer to the microscope and its parts in Figure 3.1.
3. Align the microscope as described in your notes or as outlined in the appendix of this exercise. Take note of any minor differences your instructor points out in your microscope or the procedure to be followed.
4. Starting at a lower magnification (about 40×) and then going higher (about 100×), align the microscope and observe the reference materials as the instructor directs. Record your observations and make appropriately clear drawings of what you see.
5. Carefully observe the instructor's demonstration as to how to mount a sample in an immersion medium. Immersion medium is a liquid, gas, or solid that surrounds the sample. Depending on the sample and the medium, the appearance will vary in detail.
6. Employing a 100× mount, observe loose samples that the instructor selects in at least two of the different RI mounting liquids. Record any differences you observe.

Alignment Instructions

Proper microscope alignment is essential to obtaining high-quality images and ensuring that the instrument performs to its maximum capability. Köhler illumination is a method of aligning the microscope so that the optical components are in the correct positions, distortion is minimized, illumination is bright, uniform, and concentrates on the area around the sample of interest, with extraneous glare and light being eliminated.

The components that may need adjustment are the (1) light source, (2) field diaphragm, (3) focus, (4) condenser focus, and (5) condenser diaphragm. The terms *iris* and *aperture* are often interchanged with *diaphragm*. Certain microscope manufacturers adjust certain parameters in the factory to ensure certain performance levels, so not all five components may always be adjustable. The microscope user may have to adjust the interpupillary distance on the binocular viewing head and the eyepiece focus to account for differences in diopter and allow comfortable and strainless viewing. Figure 3.1 shows the microscope components. Figure 3.2 illustrates the appearance of the fields of view for establishing Köhler illumination.

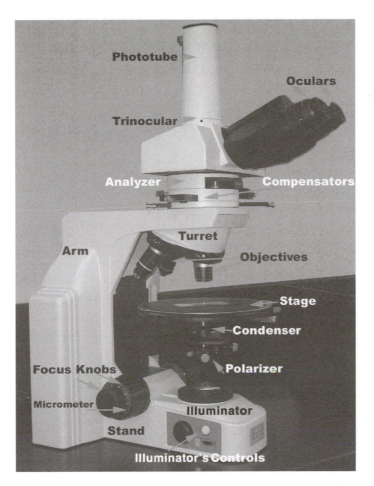

Figure 3.1
Parts of a polarized light microscope.

Köhler Illumination

1. Place the objective you intend to employ (usually 10×) into position. Turn on the illumination (if adjustable, ask your instructor about the voltage setting). Place a specimen on the stage.

2. Using the coarse focus, carefully bring the objective 1 to 2 mm above the sample by checking the space from the side of the microscope. View the specimen through the eyepieces and carefully adjust the coarse focus up (to increase the working distance) until the sample is in focus. The focus need not be perfect.

3. Adjust the interpupillary distance so that the field of view is round and you see a single image. At least one eyepiece usually has a focus adjustment. Close the eye that uses the eyepiece with the adjustment and critically focus the sample with the fine focus. Open that eye and close your other eye. Critically focus the image using only the adjustment for the eyepiece on the microscope head. The upper portion of the microscope is now properly adjusted (see Figure 3.2).

4. Close the field iris so that it can be seen in the field of view or specimen plane. Do NOT change the focus. Center the image of the iris around the center of the specimen in the center of the field of view. Focus the image of the iris by raising and lowering the condenser using its focus control. Open the field iris until it is just outside the field of view.

5. It is best if the microscope is designed so that the illuminator filament can be seen in the back focal plane or the objective. For this to be accomplished, there cannot be a frosted glass diffuser in the light path. Many modern laboratory microscopes have this diffuser, which cannot be removed, as well as precentered and focused bulbs. If this is the case, you cannot go any further; read the next paragraph and skip to the last paragraph of this section. Ask your instructor for guidance.

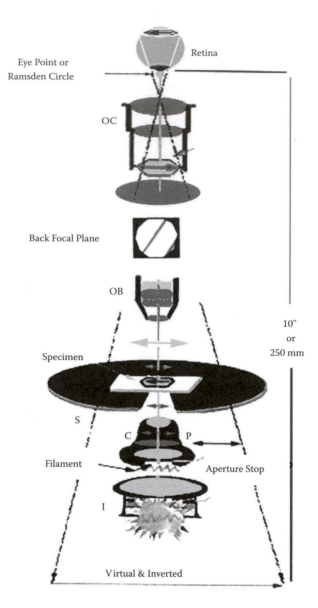

Retina

Eye Point or
Ramsden Circle

OC

Back Focal Plane

OB

10"
or
250 mm

Specimen

S

C   P

Filament   Aperture Stop

I

Virtual & Inverted

Figure 3.2

The back focal plane can be viewed in three routine ways. You can remove the eyepiece and look down the micro-scope tube. The back focal plane will be seen. By opening and closing the condenser aperture, the user can see the image of the iris. An enlarged, more easily seen image of this focal plane can be viewed if a special eyepiece, called a phase telescope or centering telescope, temporally replaces the normal eyepiece. The observer can also employ a special auxiliary lens built into the microscope as an option, called a Bertrand lens, to view the enlarged image.

Turn down the intensity of the filament so that it is dull red and not bright enough to cause discomfort to the observer. While viewing the back focal plane, use the bulb centering and focusing controls on your particular microscope to center, focus the filament, and fill the focal plane with the image of the filament.

While viewing the back focal plane, close the condenser iris until about 70 to 80% of the diameter of the focal plane view remains unblocked.

6. Restore the optical configuration of the microscope so that the specimen can be viewed properly through both eyepieces. While viewing, always make critical focus adjustments with the fine focus.

7. Perform your tasks and record your data.

Name _____ Date _____

Report

1. Label the parts of the microscope depicted in Figure 3.3.

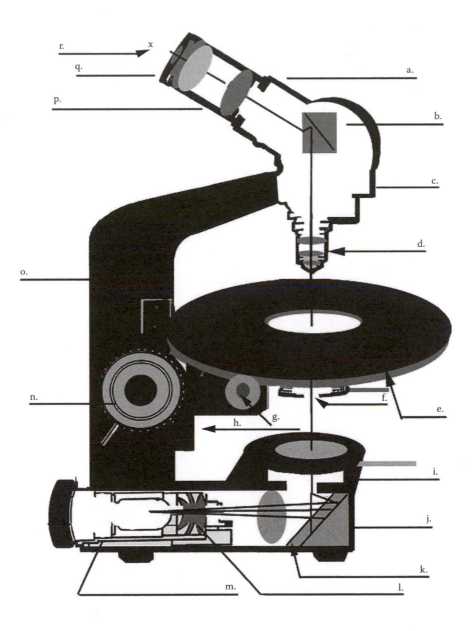

Figure 3.3

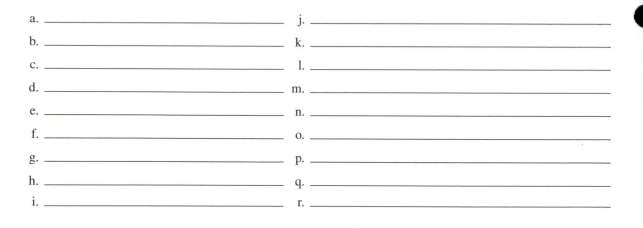

a. _____   j. _____

b. _____   k. _____

c. _____   l. _____

d. _____   m. _____

e. _____   n. _____

f. _____   o. _____

g. _____   p. _____

h. _____   q. _____

i. _____   r. _____

2. Use the space below to write a report of your observations; support your findings with the drawings you made (see Hands-on Exercise, step 4).

4

Stereomicroscopes and Firing Pin Impressions (Tool Marks)

Teaching Goals

This experiment is divided into several parts:

- The student will learn about the usefulness of the stereomicroscope (dissecting microscope).
- The student will learn how to calibrate an eyepiece scale (ocular micrometer) with a standard.
- The student will learn how to use the ocular micrometer to perform a number of measurements.
- The student will learn how to employ the stereomicroscope to observe and measure the surface characteristics of various frequently encountered items of evidence in an effort to establish a common origin for two or more items.

Background Knowledge

The stereomicroscope (SM) is a crucial piece of equipment in the forensic laboratory. It is used by forensic scientists for the examination of all types of physical evidence. Forensic biologists use the SM to examine potential blood and seminal stains; forensic trace evidence examiners use the SM to examine glass fragments, paint chips, hairs, fibers, textiles, ropes, broken objects, and soil, to mention just a few; forensic anthropologists use the SM to examine skeletal remains; odontologists use the SM to examine bite marks; and forensic archeologists use the SM to examine all manner of artifacts. Therefore, it is not surprising that firearms and tool mark examiners frequently make use of the SM in their preliminary examinations of firearms, bullets, projectiles, cartridges, shell casings, ballistics, and obliterated numbers. This type of microscope is also very useful in the preparation of samples for further analysis with other techniques such as x-ray diffraction, microspectrophotometry, and scanning electron microscopy. The primary reason the SM is used for so many applications in forensic science laboratories is that it produces a moderately magnified, three-dimensional, macroscopic image, thereby allowing for the close-up examination of most forms and types of physical evidence.

The SM is often used to assist in making measurements of macro-size items of trace evidence. In order to accomplish this goal, the microscope is equipped with an eyepiece into which has been installed an ocular micrometer on which there is a measurement scale. This scale needs to be calibrated so that the analyst knows the projected dimension of each division of the eyepiece scale at each magnification the analyst wishes to use for measurements. This scale is calibrated with a standard scale known as a stage micrometer. A stage micrometer is a known linear length, such as 1 mm divided into 100 equal lengths or divisions. Therefore, each division on the stage micrometer scale is equal to 10 μm.

Equipment and Supplies

1. Stereomicroscope, zoom preferred approximately 5× to 40× or 10× and 20× dual magnification with reflected light illuminator. Eyepiece equipped with an ocular scale is a preferred option.
2. A reflected light stage scale or a millimeter scale to be used for ocular scale calibration.
3. Index cards (light blue or green in color), glass or disposable petri dishes 75 to 125 mm in diameter, scalpel, dissection needles (minimum of two), microspatula, small scissors, and tweezers.
4. Selected items for examination, such as type writing, photocopy, color printing from a magazine, paper currency, a fired bullet, a sample of dust and debris.
5. A prepared set of at least three discharged cartridge cases, two of which may or may not have been fired from the same firearm.

Hands-on Exercise

Part I: The Stereomicroscope

Review the diagram of the stereomicroscope in Figure 4.1. Note that there are two distinct optical systems viewing the sample at approximately a 15-degree angle of divergence. These two optical paths enable the stereomicroscope to produce a three-dimensional appearance of samples.

The lab instructor will explain how to turn on the illuminator (be careful not to overpower the bulb), focus the microscope, adjust for your interpupillary distance, and correct for differences in the diopter of your left and right eyes.

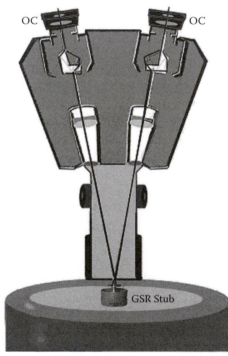

Figure 4.1
A schematic of a stereomicroscope showing two optical systems.

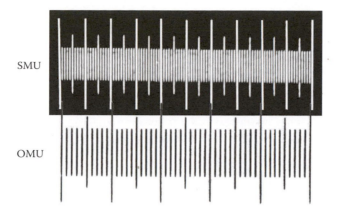

Figure 4.2
The superimposed image of the stage micrometer is aligned with the image of the ocular micrometer.

Optional Exercise

If your microscope is equipped with an eyepiece scale, you need to calibrate the dimension of the projected eyepiece scale at a low (10×) and high magnification (30×). This is accomplished by viewing the standard (stage micrometer or quality millimeter scale) under the microscope and aligning the eyepiece scale with the standard scale (see Figure 4.2). Count the number of stage micrometer units (SMUs) and the number of ocular micrometer units (OMUs) that occur between two aligned scale marks on each scale. Make a table similar to that below (Table 4.1) in your notebook and fill in the data. Calculate the projected dimension of each eyepiece scale unit at the desired magnification. The collected data should be recorded in Table 4.1.

TABLE 4.1
Calibration of an Ocular Micrometer using a 1 mm Stage Micrometer Divided into 100 Equal SMUs (the OMU must be calibrated for each magnification setting on the SM)

Magnification Setting	No. of SMUs	No. of OMUs	1 SMU =10 μm	1 OMU = $\dfrac{\text{No. of SMUs} \times 1\ \text{SMU Value}}{\text{No. of SMUs} \times \text{No. of OMUs}}$

The superimposed image of the stage micrometer is aligned with the image of the ocular micrometer. If the stage micrometer is 1 mm divided into 100 equal divisions, then each SMU is equal to 10 μm. The ocular micrometer is divided into 50 equal divisions. The value for each OMU is determined by calibration with the stage micrometer and the following formula:

$$1\ \text{OMU} = \text{No. of SMUs} \times 1\ \text{SMU value}/\text{No. of SMUs} \times \text{No. of OMUs}$$

Part II: Observations with the Stereomicroscope

The instructor will supply you with a number of items to view with the microscope. If available, view at both low and high magnification settings. Record all observations in your notebook. The inclusion of a sketch or drawing (labeled) may make your recording easier and more meaningful. If you have calibrated your eyepiece micrometer, it would be appropriate to include measurements with your observations.

Using the discharged cartridge casings or some other item that has a textured surface, observe the item with oblique lighting originating from different directions and axial lighting (straight down) pointed directly into the object being observed. Oblique lighting can be accomplished by keeping the light stationary and rotating the sample or by keeping the sample in one position and moving the light source around the periphery while the angle of the light is varied from the horizontal. At least once, the illuminator should be placed at a very high angle, almost vertical to the surface observed. Record all your observations.

Part III: Comparison of Firing Pin Impressions

You will be supplied with at least three discharged cartridge casings. Two or more of these may have been fired from the same weapon. They will have some identifying mark on each. View each under the microscope at an appropriate and comfortable magnification. Record your observations. A drawing may be appropriate. Measurements may also be appropriate. Based on your observations, determine if any of the casings were fired from the same weapon.

You should keep in mind that these observations and determinations are in reality a specific form of tool mark examination and the same principles of identification of tool to substrate apply (see Figure 4.3 and Figure 4.4).

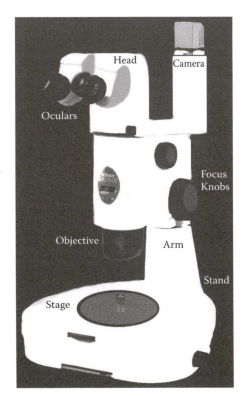

Figure 4.3
Stereomicroscope with shell casing.

Figure 4.4
Macrophotograph of the firing pin impression with the image of the ocular micrometer superimposed over the primer cap.

Name _____ Date _____

Report

1. Report the results of the eyepiece scale calibration if performed.

2. Report the results of your cartridge case comparison.

3. Discuss in short narrative what you have learned in this experiment.

4. How does the image obtained with the stereomicroscope differ from that of the image obtained with the compound microscope?

Acquiring and Classifying Inked and Latent Fingerprints

Teaching Goals

1. Each student will learn how to obtain an inked fingerprint set from another member of the class.
2. Each student will then classify each print according to their major group of classification.
3. Each student will learn how law enforcement personnel obtain reference fingerprints and classify them for future retrieval and identification of persons.
4. Each student will learn how to locate, develop, collect, and preserve latent (invisible) fingerprints by powder dusting and tape lifting the hidden print.
5. Each student will learn how crime scene personnel or investigators locate and preserve invisible ridge impressions left behind at a crime scene.

Background Knowledge

The friction ridges covering the tips of our fingers, the surfaces comprising the palms of our hands, and the outer surfaces of the soles of our feet and toes are formed during gestation. These ridges are fully developed at the time of birth. The patterns formed by these ridges are known to stay the same throughout one's life. In addition, it is believed that no two people have the same or identical fingerprints. Consequently, the ridge patterns present on our fingers, palms, soles, and toes have been used for identifying human beings for more than a century. It is believed that the variables in the ridge characteristics and patterns make it statistically certain that no two individuals alive will have identical prints. This idea is so ingrained in our collective psyche and culture that the sole patterns of all newborn infants' feet are inked and recorded in the delivery room by one of the delivery room nurses minutes after birth.

Of the four areas containing friction ridge patterns on the human body, fingerprints have received the most attention in the areas of positive identification and criminal investigation. The patterns of the friction ridges of the fingertips are readily transferred when they contact a surface. Secretions from the pores of the skin (sweat), composed of approximately 99% water and 1% inorganic and organic molecules, are readily transferred from one's fingertips to any object they touch.

The various categories of fingerprints that can be found at a crime scene are invisible or latent prints (invisible deposit of sweat or body oils on a surface), visible or residual prints (opaque deposits of blood, dirt, soot, or other materials transferred to a surface from the tips of the fingers), and plastic or patent prints (impressions of the fingerprints in soft materials such as putty or clay).

Latent or invisible fingerprints are made visible through the use of powders, chemicals, or other means. Visible fingerprints are observed by eye and/or with low magnification (5× to 10×). They are made when fingers are covered with an opaque substance such as soot, grease, blood, or ink that is transferred onto a

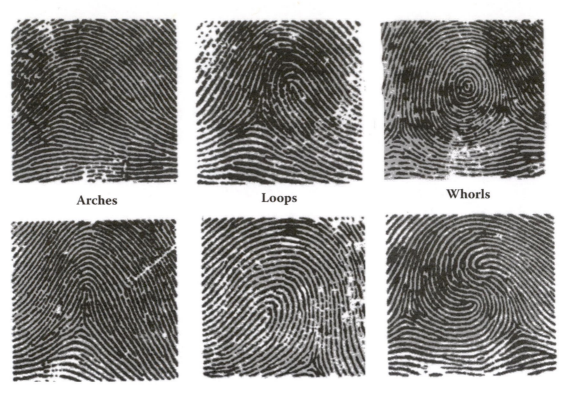

Figure 5.1
The three basic categories of human fingerprint patterns: arches, loops, and whorls.

surface when contact occurs between the fingertip and the surface. Plastic prints are formed when fingers come in contact with a soft, pliable material that can take an impression, such as putty, wax, wet paint, etc. Plastic prints are reversed images of the fingerprint ridge patterns and thus can be cast with silicone casting material.

There are three basic fingerprint configurations (see Figure 5.1):

Loops — Loop patterns are classified as either ulnar or radial and have one delta.
Arches — Arch patterns are classified as either plain or tented.
Whorls — Whorl patterns are classified as either plain, double, central pocket, or accidental and have two deltas.

Equipment and Supplies

1. Fingerprint ink and roller with either a piece of glass (approximately 18 × 9 in.) or other smooth surface to spread the ink.
2. Fingerprint cards (Appendix A), magnifying glass (10× or greater) or professional fingerprint examination glass with scale.
3. Glass or ceramic items (beakers or coffee cups) to dust for latents.
4. Dark and light fingerprint powders or multipurpose powder with dusting brushes (at least two; one for each color).
5. Latent fingerprint lifting tape with backing material or hinged lifters (dark and light).

6. Fingerprint ink remover towelettes or paper towels with waterless hand cleaner or denatured ethyl alcohol.

7. A fingerprint dust safety hood for containing fine dust particles (optional).

8. Fingerprint magnifier.

9. Camera to photograph prints with 1:1 (life size) ability, if possible. Life size means that the image on the film is the actual size of the object photographed.

Hands-on Exercise

1. Review the section on fingerprint classification in your textbook.

2. Obtain blank fingerprint cards, inking materials, and magnifier from the instructor.

3. Prepare two sets of fingerprint cards. Clearly print the name of the person who is being printed as well as the person collecting the print on one of the fingerprint cards and leave one card blank (without identification). The collector will be graded on the quality of the prints submitted. Both students will be graded on their classification of the prints.

4. Collect a set of rolled and plain inked prints from a fellow student.

5. Follow the instructions given by the instructor. We have found that students usually obtain better results if the printer always stands to the subject's **left** no matter which hand's fingers are being rolled. The subject should stand behind the printer about a forearm's length from the table. Follow the directions carefully as to which way, toward or away from the subject's body, the prints should be rolled. Keep in mind that usually prints are ruined because **too much** ink is used rather than too little. Be sure to roll the ink out very well prior to use.

6. After the printing is complete and clean up of fingers and the printing station is complete, the printer should classify the prints according to the instructor's directions. This is usually just the major classification, but the instructor may require more. Use the examples in Figure 5.1, Appendix B, and your textbook. Record these results in your notebook and in the proper place on the card; the printer first and the person printed in brackets, for example, (Student 1) [Student 2].

7. Obtain a latent dusting kit from the instructor. Make a few latent prints on a dark, smooth surface, a light (white or ivory) surface, and on clear glass, if all are available. Use a different finger for each surface. If you are unsuccessful developing a print (not making it visible), it is likely because your fingers are clean and dry (see Figure 5.2). Repeat the exercise after touching your fingers to the side of your nose.

8. Choose an appropriate contrasting fingerprint dusting powder and the proper brush for its application. **DO NOT** interchange or reuse brushes with different colors and types of fingerprint powders, as this will result in poor results. Tap or twirl the brush to remove any excess powder left from the last use.

9. Place a small amount of the powder into the jar cover or onto a piece of **disposable** paper (not onto your notebook). Dip the tip of the bristles into the powder. It is considered poor practice to have powder up into the center of the brush bristles, as this causes too much powder to be deposited and will ruin an otherwise good print. Using a circular sweeping motion, **lightly** brush across the area where the latent print is suspected of being. Concentrate your action on the area where the print develops (becomes visible). Pick up a small amount of additional powder if required. Stop when development is complete, too much brushing can destroy a print. Carefully remove excess powder with a clean brush.

10. Photograph the print, preferably 1:1 or life size (optional).

11. Follow the directions from the hinged lifter supply company to lift the print. The following are general directions:

    a. Open the lifter so that the arrow points to the upper right. Starting at the upper right, peel off the plastic cover and discard it. The adhesive is now exposed.

    b. Place the tacky side of the lifter carefully over the latent image or print. Rub the entire plastic sheet completely, but lightly. Lift the powdered print. Note that the ❶ symbol is facing you when lifting the print.

    c. The cover of the hinged lifter now protects the lifted print from scratches, damage, and dirt. To complete the lift, cover the lifted print by placing the lifter on its back (that is with the adhesive side up), then form a gentle curl with the backing cover and carefully roll the cover over the print lifter.

Figure 5.2
The use of a feather duster with white powder employed to develop a latent fingerprint on a dark object.

    d. The lifted print is now sealed. An attempt to separate the backing and the lifter will result in damage or destruction of the lifted image. Keep the ❶ facing you (face up), this will ensure that the print is a positive image, that is, it will appear as it did on the surface on which it was found (see Figure 5.3)

12. Classify the lifted prints to a minimum of the major group. Go further if the instructor wishes. Record your results in your notebook and tape the lifts under the results.

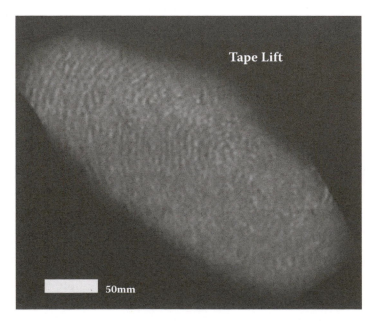

Figure 5.3
Latent print lifted with tape and placed on a black background.

13. Turn in the inked print card to the instructor. Photocopies may be made of it. The original or a copy will be returned at a later time. Leave a space (page) in your notebook to paste the card.

14. Clean the work area, removing all dusting powder and ink residue. Return materials to the instructor or to a predetermined area in the lab.

Name _____ Date _____

Report

1. In a short narrative, discuss what you have learned in this exercise; include the skill you gained in preparing inked prints and developing latent prints with lifting techniques.

Appendix A

Identity of person printed _____ Date _____

Identity of person doing collection _____

Classification according to loops, whorls, and arches _____

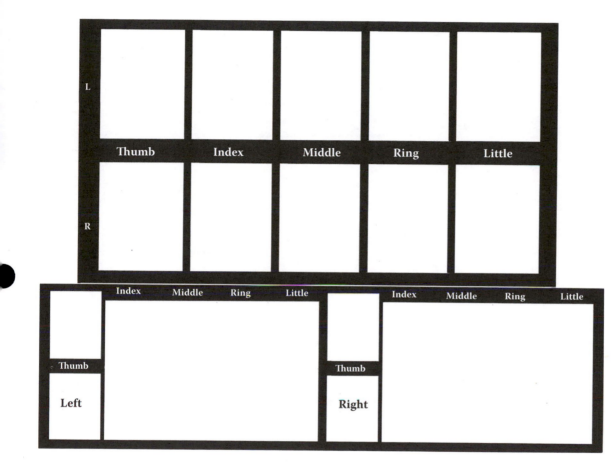

Figure 5.4

Appendix B

A Few Fingerprint Characteristics

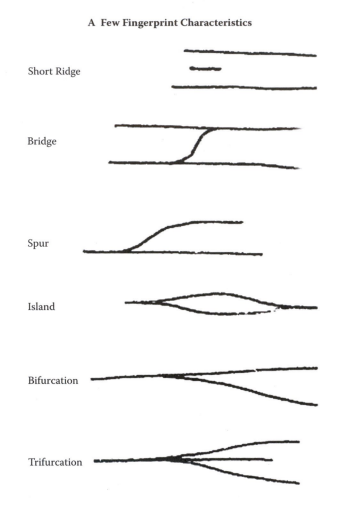

Short Ridge

Bridge

Spur

Island

Bifurcation

Trifurcation

Figure 5.5

Experiment 6

Identification and Matching of Fingerprints

Teaching Goals

Upon completion of this exercise each student will be able to examine and compare questioned and known fingerprints. In addition, each student will know the criteria necessary for the identification of individuals from their fingerprints.

Background Knowledge

Today, even with the advent of DNA analysis, fingerprints still remain the most positive method of identifying individuals. Fingerprints are impressions made by the friction ridges present on the outer surfaces of the distal ends (tips) of the fingers. Friction ridges begin to develop in the third or fourth month of gestation. At birth, an individual is born with a fully developed set of friction ridge patterns that remain constant throughout his or her entire life.

Fingerprints consist of class patterns: arches, whorls, and loops (see Experiment 5). Pattern types are class characteristics sufficient for placing individuals into broad groups; however, they are not very useful for identification and individualization purposes.

Equipment and Supplies

1. A set of "known" fingerprint cards, prepared in Experiment 5, to use as standards for comparison and identification of prints.
2. A set of unknown fingerprints.
3. Magnifying glass.

Hands-on Exercise

1. Obtain a set of "known" reference prints from your instructor. The identity of the prints can be found on the card.
2. Obtain a set of unknown prints from the instructor.
3. Record the unknown envelope number in your lab book.
4. Compare each unknown found in your envelope to the known cards. All prints need to be identified or reported as "No reference found." For identifying and individualizing features, see Figure 6.1 to Figure 6.4.
5. Record the unknown identifier (G-4) for each unknown in your lab book and the identity from the known card (if there is a match) and the number for the finger that appears on the card for the match. Do this for all the unknowns in the envelope.
6. Return the knowns and unknowns (in their envelope) to the instructor.
7. Hand in your notebook as requested by the instructor. You may not be late with this assignment.

Some Important Features and Their Definitions

Delta — A delta is the place on a ridge nearest to the point of divergence of two type lines.

Core — A core is the center of a loop, whorl and some arch patterns.

Type Line — Type lines are two ridges that run parallel and surround or tend to encircle the pattern area.

Friction Ridge Line — Friction ridges are the crests or high points of the lines present on the underside of our hands and feet. These ridges are abundant and form unique and individualizing patterns on the outer surfaces of the skin of our fingers, palms, toes, and soles. The troughs or low points of these ridges are known as furrows.

Bifurcation — Bifurcations are the forking of two friction ridges.

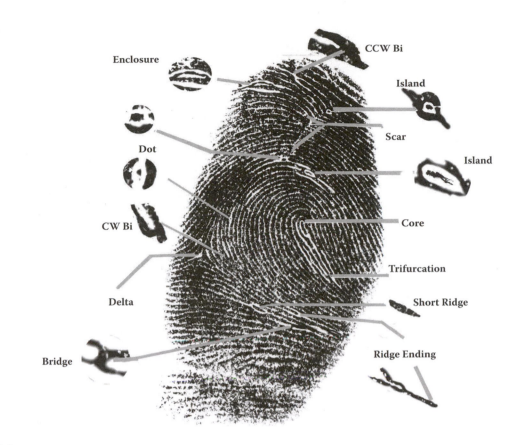

Figure 6.1

Some of the individual ridge features found in human fingerprints used in positive identification or elimination of an individual as being the source of a particular fingerprint. The figure contains examples of many of the individual ridge features found in fingerprints. These features in relationship to one another, along with class pattern (see Experiment 5) and ridge counts, are used in fingerprint identification. In addition, scars and injuries are also useful (see Figure 6.3).

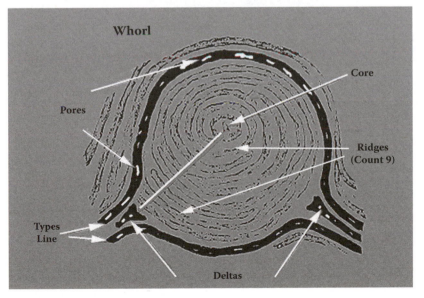

Figure 6.2

Examples of classification features and ridge count.

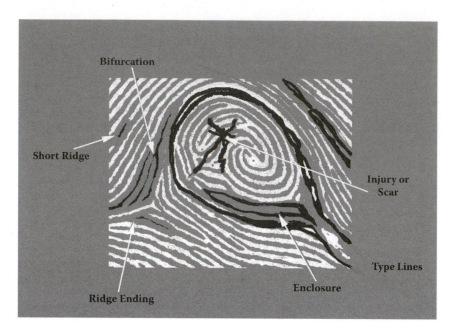

Figure 6.3
Example of scarring or injury.

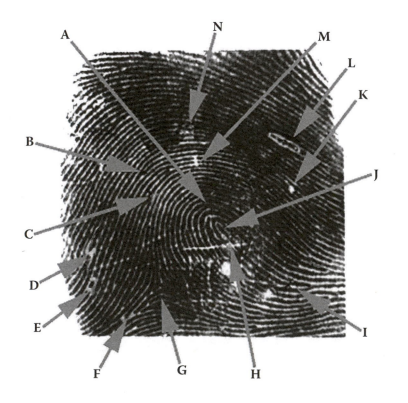

A- CCW bifurcation
B- CW bifurcation
C- CCW bifurcation
D- Ridge ending
E- Dot
F- Ridge ending
G- Delta
H- Scar
I- Enclosure
J- Core
K- CW bifurcation
L- Short ride
M- Scar
N-Dot

Figure 6.4
Fingerprint chart with identifying features. What is the class category of this fingerprint? Can you find any more identifying features?

Name _____   Date _____

Report

1. In your notebook, record the identification numbers and/or letters of the unknown fingerprints and next to them record the identification numbers and/or letters of the reference or known fingerprints that match the unknown.

2. Make a list of at least 14 matching features between the questioned and known fingerprints. Figure 6.4 is an example of a fingerprint chart showing the identifying features used for positive identification.

1. _____   2. _____

3. _____   4. _____

5. _____   6. _____

7. _____   8. _____

9. _____   10. _____

11. _____   12. _____

13. _____   14. _____

Cyanoacrylate "Super Glue" Fuming Technique to Develop Latent Fingerprints

Teaching Goals

The student will learn how to develop latent fingerprints on a plastic surface by the "Super Glue" fuming method.

Background Knowledge

Chemical fuming has a long history of use in latent fingerprint development. For nearly a century, forensic investigators have been using the fumes coming off warmed crystals of iodine to develop latent fingerprints on paper. Normally the procedure involves hanging the paper in a closed fuming chamber containing a small tray of iodine crystals. The chamber is closed and the iodine crystals are gently heated using a hot plate. Upon warming, the solid crystals of iodine sublime, or go directly into the vapor phase, and saturate the environment in the small chamber. The iodine fumes then interact with the invisible or latent fingerprint residues present on the paper and form brownish-colored visible prints.

Similarly, cyanoacrylate (or Super Glue) fumes have been used to develop latent prints on objects made from plastics and other hard to print surfaces. In the late 1970s, forensic investigators in Japan first developed the method of using cyanoacrylate fumes to find latent fingerprints. Since then, much research has been done by the forensic science community to improve the original procedure. This exercise introduces the use of Super Glue in fingerprint development.

Equipment and Supplies

1. A small item (provided by student) to be used for fuming with cyanoacrylate glue
2. A set of unknown fingerprints
3. Fuming chamber (**MUST BE USED UNDER A FUME HOOD)***
4. Super Glue
5. One small warming plate
6. One disposable aluminum cup
7. One 100 mL beaker with water
8. One piece of clear plastic (standard)

---

* An inexpensive chamber can be made from 4-mm thick plastic sheets.

9. Red fluorescent dusting powder

10. One fiberglass dusting brush

11. One magnifying glass

Hands-on Exercise

1. The item to be fumed is hung or placed in the fuming chamber along with a standard piece of plastic. The fuming chamber must be located under the exhaust hood or be hooked up to an external ventilation system (see Figure 7.1).

2. The 100 mL beaker or flask containing 50 mL of water is placed into the chamber.

3. The cup warming plate is placed on the bottom of the fuming chamber.

4. A clean aluminum cup is also placed into the chamber on top of the warming plate.

5. Pour approximately 2 mL of glue into the aluminum dish.

6. Close the door to the chamber as soon as possible. (**FUMES ARE HAZADOUS IF INHALED. GOOD EXTERNAL VENTILATION IS ESSENTIAL.**)

7. Allow the fuming process to proceed for 45 min.

8. While fuming the specimen, draw a sketch of the apparatus.

9. Open the door to the chamber slowly; this will allow for the cyanoacrylate fumes to be drawn up by the exhaust hood.

10. Before opening the fume hood to examine the item you must have all safety equipment on (**IN PARTICULAR, MAKE SURE YOU ARE WEARING YOUR SAFETY GLASSES**). Examine the impression with ambient light; next, use oblique lighting to examine each impression. Describe the differences in what you observe using each lighting technique.

11. Photograph the latent print.

12. Dust the print with orange fluorescent dusting powder. Examine under an ultraviolet (UV) lamp. **CAUTION!!! NEVER LOOK INTO A UV LAMP; THE UV RADIATION WILL BURN YOUR EYES!!!!!!!!**

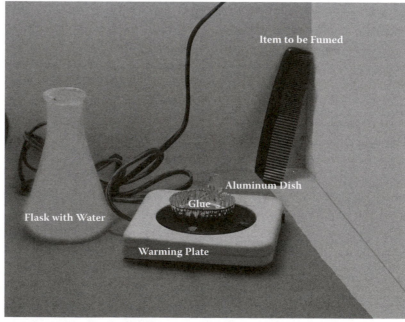

Figure 7.1
Fuming chamber for the "Super Glue" process.

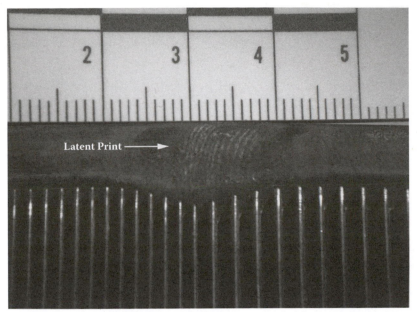

Figure 7.2
Fumed print on a comb.

Name _____ Date _____

Report

1. Draw a sketch of the apparatus.

2. Record the results as requested by your instructor.

3. If possible, classify the latent print.

8

Crime Scene Investigation
Safeguarding, Searching, Recognition, Documentation, Collection, Packaging, and Preservation of Physical Evidence Found at the Crime Scene

Teaching Goals

After this lesson, each student will know the procedures for the safeguarding, searching, recognition, documentation, collection, packaging, and preservation of most categories of physical evidence recovered at the scene of a crime or other scientific inquiry.

Background Knowledge

Systematic plans for the searching, recognition, documentation, collection, packaging, and transportation of physical evidence recovered at the location of any scientific inquiry are essential for the accurate reconstruction of past events. Crime scenes, in particular, must be meticulously processed and documented. Any chance for contamination of and/or tampering with the evidence must be prevented, for the very freedom and/or life of the accused may be at stake. Whether the event is a crime that happened hours ago or an archeological site from past millennia, a thorough, systematic examination of the scene is essential for elucidating the truth and reconstructing as accurately as possible what actually occurred or existed at the time of the event.

Equipment and Supplies

1. A prepared mock crime scene
2. Diagram of a crime scene
3. Video camera with tape
4. Still (35 mm) camera, film, flash, and other required accessories
5. Measuring devices: rulers, tape measurers, protractors, compass
6. Protective clothing
7. Crime scene tape
8. Evidence searching equipment
9. Evidence collection and packaging equipment

Hands-on Exercise

One student will be chosen for the role of first officer at the scene. Two students will be chosen to be the two investigation detectives. Two students will be chosen as emergency medical personnel. One student will be chosen to be the medical examiner. Two students will be chosen to be crime scene technicians assigned to process the crime scene. One student will be chosen to be the forensic scientist called to the scene to assist with processing. Two students will act as a reporter and cameraperson for the news media. The remaining students will act as witnesses, nonessential police personnel, and bystanders.

1. Based on the requirements of the role assigned to you, discuss the procedure you would follow at the crime scene shown in Figure 8.1. For example, emergency medical personnel would be concerned with the victim, not with processing evidence; a reporter might want to talk to the medical examiner, etc.
2. The instructor will direct an interactive dialogue with students in their assigned roles and discuss essential aspects of recognizing, finding, collecting, safeguarding, documenting, packaging, and preserving the evidence found at crime scenes.
3. Process the prepared crime scene.

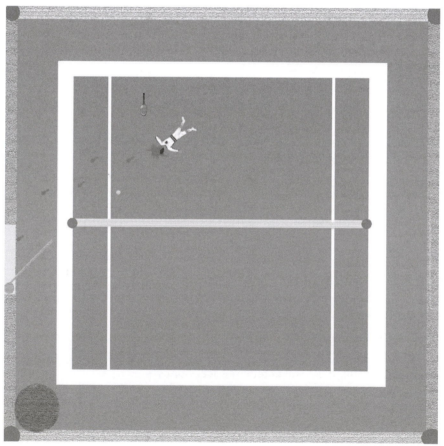

Figure 8.1
The prepared crime scene.

Important Concepts and Observations

- Condition of the victim; documentation
- Medical treatment arranged if necessary
- Note taking: who, what, when, where, why, how
- Searching techniques: walk-through, walk-out
- Search patterns: spiral, grid, wheel, zone
- Measuring techniques: baseline, triangulation, polar coordinate, fixed point
- Crime scene documentation
- Crime scene diagram (including key, scale, legend, conditions, etc.)
- Crime scene security
- Exit of nonessential personnel, bystanders, etc.
- Evidence handling

Crime Scene Activities

1. Walk-through — A prompt and moderately slow walk-through the crime scene by assigned officers. The walk-through starts at the outer limits of the crime scene and proceeds toward the seat of the crime. The intent is to obtain a quick overview of the physical evidence (see Figure 8.2).

2. Suggested methods of searching for physical evidence at crime scenes:

   a. Spiral method: Start the walk-through at the outer border of the scene and continue in a spiral pattern toward the seat of the crime, or start at the center of the spiral (seat) and proceed to the outer limits of the scene.

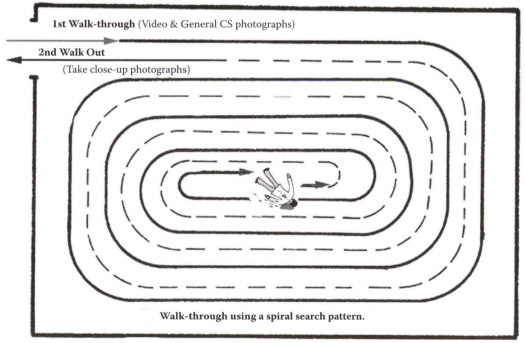

1st Walk-through (Video & General CS photographs)

2nd Walk Out

(Take close-up photographs)

Walk-through using a spiral search pattern.

Figure 8.2

An initial, prompt, slow walk-through of the entire crime scene should be performed. A similarly slow walk should be conducted when exiting the scene for the first time, thus making sure that no vital evidence is missed.

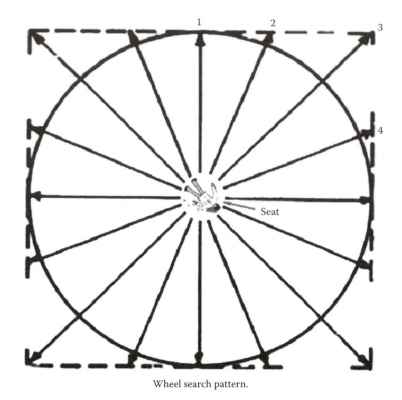

Wheel search pattern.

Figure 8.3
Search from the seat of the crime outward along a radial path starting from the 12 o'clock position and moving clockwise around the scene.

    b.   Wheel method: Start at the seat of the crime and walk away from the seat along equal-length radial lines. Return to the center via the same radial line (see Figure 8.3).

    c.   Zonal method: Divide the crime scene into small sectors or zones and search each one individually. This method is effective for large area searches or vehicle searches (see Figure 8.4).

3. All physical evidence located during a search should be vigilantly protected, documented, and collected by trained personnel.

4. Documentation of the crime scene and physical evidence:

    a.   The pristine crime scene should be recorded by the videographer who should accompany an assigned officer on the initial walk-through.

    b.   Overall and specific photographs should be taken of the crime scene.

    c.   Close-up (examination quality) photographs should be taken of important items of evidence on the ground and walls (e.g., tire impressions, footwear impressions, blood stain patterns, etc.). The largest photographic format available should be employed for closeups. The print image should fill the entire frame. A ruler should appear in the photograph to clearly show scale. A tripod should be used to hold the camera in a horizontal position. A level should be available to help properly position the camera.

    d.   Precise measurements of the crime scene and descriptive notes must be made. The notes should include an accurate sketch containing a key, a scale, and a legend noting the date, time, location, and conditions (weather, lighting, etc.). Compass directions should be noted on the sketch (see Figure 8.5 and Figure 8.6).

    e.   A formal sketch should be prepared as soon as practical from the rough sketch and descriptive notes generated at the scene.

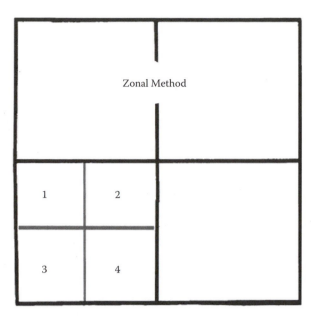

Figure 8.4
The entire area is divided into distinct zones. The zone adjacent to the seat of the event is searched first. Subsequent zones are searched in order of their perceived importance.

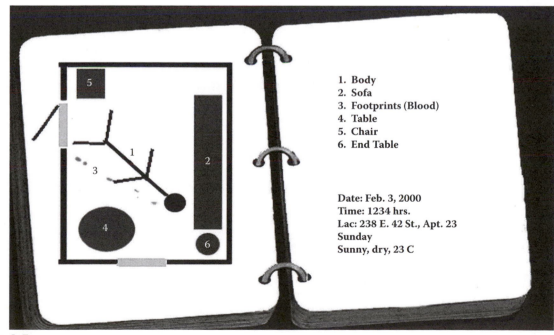

1. **Body**
2. **Sofa**
3. **Footprints (Blood)**
4. **Table**
5. **Chair**
6. **End Table**

Date: Feb. 3, 2000
Time: 1234 hrs.
Lac: 238 E. 42 St., Apt. 23
Sunday
Sunny, dry, 23 C

Figure 8.5
Investigator's notebook with notes and a rough sketch.

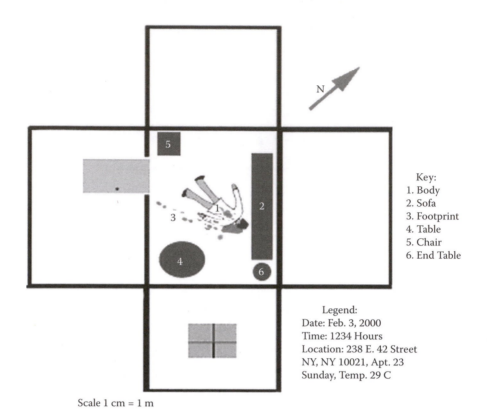

Key:
1. Body
2. Sofa
3. Footprint
4. Table
5. Chair
6. End Table

Legend:
Date: Feb. 3, 2000
Time: 1234 Hours
Location: 238 E. 42 Street
NY, NY 10021, Apt. 23
Sunday, Temp. 29 C

Scale 1 cm = 1 m

Figure 8.6
A formal sketch of the crime scene with a scale, legend, key, and direction (North) all incorporated into the drawing.

Name _____ Date _____

Report

1. List the procedures for processing the prepared crime scene. Note all dos and don'ts.

2. Write an extensive report discussing all the details concerning processing of the prepared crime scene. The report should be prepared as if you were going to testify in court.

3. Write a formal critique as to what you think was done right at the crime scene, what you think was done improperly at the crime scene, and, finally, how you think the crime scene should have been processed.

Trace Evidence Collection and Sorting

Teaching Goals

1. In the first portion of this exercise, students will learn how to collect trace materials obtained from items of clothing.
2. Students will learn how to package the evidence, seal it, and mark it for identification.
3. In the second part of this exercise, students will learn how to examine tape lifts under a stereomicroscope and remove the trace evidence for examination.
4. Finally, students will learn the concept of **chain of custody** and why it is so important in our legal system.

Background Knowledge

Trace evidence in the form of small particles of matter and tiny bits of fibrous materials left at crime scenes or transferred between the people, places, and things involved in a crime have been used extensively in solving crimes and reconstructing events. Thus the recognition, documentation, collection, packaging, and preservation of trace evidence is of vital importance in all forensic investigations. Trace evidence is collected in many ways. However, a systematic approach to the collection of trace evidence is recommended.

Typically the article to be examined is removed from its package (if possible, the security seal made by the packaging technician should not be broken) and placed on a clean examination table covered with fresh examination paper. The item is examined visually for any trace materials. Any visible traces are gently removed with fine tweezers or by scraping with a spatula. The collected materials should be sorted and placed in separate paper envelopes. The article's surfaces should then be taped to remove any trace material not previously collected. The tapes are secured by placing them on new, transparent sheets of heavy duty Mylar. Finally, tiny pieces of substrate material are removed from the item for use as known standards.

Equipment and Supplies

1. One roll of 2-in. lifting tape
2. Several pieces of clear plastic
3. One pair of tweezers
4. Packaging containers
5. One stereomicroscope
6. One magnifying glass
7. One UV light [**CAUTION! NEVER LOOK INTO A UV LAMP; THE UV RADIATION WILL BURN YOUR EYES!!!**]
8. Items of clothing supplied by the instructor

9. Large clean sheets of paper
10. Large table
11. Evidence storage cabinet
12. Sorting dishes

Hands-on Exercise

1. Clean the table with water and paper towels.
2. Spread a sheet of clean examination paper on top of the examination table.
3. Remove the item of clothing from the evidence bag and place it on the prepared tabletop; avoid breaking any security seals.
4. Make a sketch of the item of clothing in your notebook.
5. Examine the item of clothing with your eyes. Remove any visible fibrous materials and place in a paper envelope as demonstrated in Figure 9.1. Document the location of the visible trace evidence in your notebook.
6. Examine the clothing with a UV lamp (**USE CAUTION!**). After noting the location of the fluorescing fibrous materials in your notebook, remove and place them in a paper envelope.
7. Before taping the item, place a sheet of clear plastic next to the clothing.
8. Remove a 6-in. piece of clear fingerprint tape and gently tape the surfaces of the item of clothing, as shown in Figure 9.2.
9. Place the adhesive side of the tape down onto the clear plastic sheet (it should adhere to the plastic) (see Figure 9.3).
10. After placing identifying notations onto each prepared evidence package (Figure 9.4), place them all into an appropriate size envelope.
11. Place identifying information onto the evidence envelope, and seal, date, and initial.

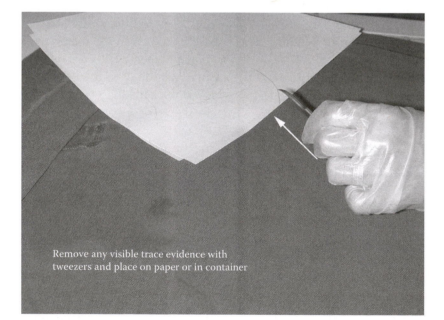

Remove any visible trace evidence with tweezers and place on paper or in container

Figure 9.1
Removing visible trace evidence with tweezers.

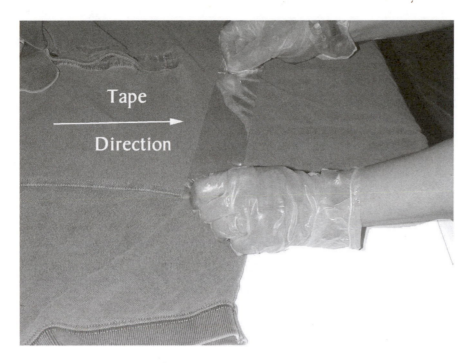

Figure 9.2
Taping a garment for trace evidence.

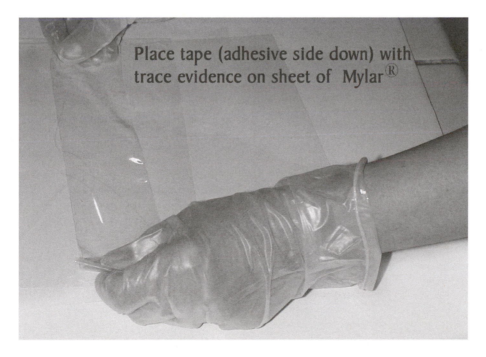

Figure 9.3
Place tape adhesive side down onto Mylar sheet to secure trace evidence removed from the garment and protect it from contamination.

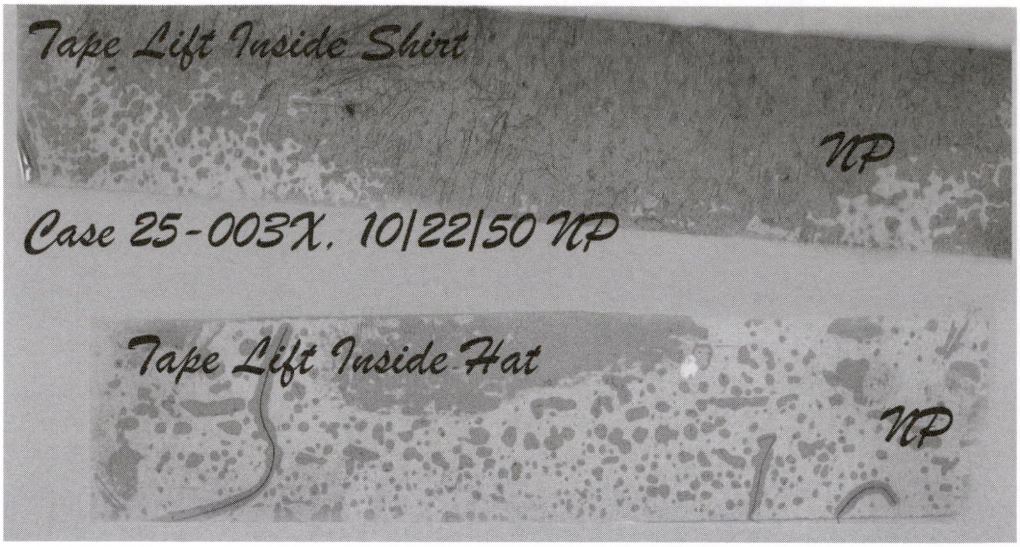

Figure 9.4
Marking tape lifts for identification.

12. Open the evidence packages and make notations as to their conditions and what you find inside each envelope (you will not get the envelope you prepared).

13. Each package must be described in great detail. All relevant information must be clearly marked in your notebook.

14. The items of evidence are opened one at a time and examined under the stereomicroscope. The fibrous materials are sorted by type. All hairs are placed in one container and all fibers are placed in a separate container. All items are sealed, marked for identification, and placed in the evidence cabinet. Fill out a chain of custody form (Appendix A) for each package of evidence.

Name _____ Date _____

Report

   1. Sketch the item of clothing you examined.

   2. Write a complete report in your notebook outlining your examination and observations. On your sketch of the clothing item, note the area where the fibers were found during your visual examination.

   3. Why do you think it is important to mark the location of the fibrous evidence before it is removed from the clothing?

4. Why is it important to mark for identification and seal the evidence?

5. Describe the chain of custody that was developed in this exercise.

6. Why is chain of custody important?

APPENDIX A

Chain of Custody Form

Evidence control no. _____

Evidence received from _____

Time _____ Date _____

Evidence received by _____

Time _____ Date _____

Sample Preparation for Microscopic Examination

Teaching Goals

Upon completion of this exercise each student will be able to prepare various types of trace evidence for examination and comparison with the light microscope.

Background Knowledge

Microscopic-size items of trace evidence in the form of hairs, synthetic fibers, natural fibers, and small particles are frequently encountered in forensic investigations. These types of minuscule items of trace evidence are easily transferred during the commission of a crime. Thus these items are frequently used to associate the people, places, and things involved in crimes. Most, if not all, of these specimens must be examined with a light microscope. However, prior to examination with the light microscope, these specimens must be properly prepared. The following exercise is designed to teach the novice forensic scientist several of the methods used in the preparation of microscope specimens prior to microscopic examination.

Equipment

1. Known tufts of hair and fiber specimens
2. Glass microscope slides and cover glass
3. Assorted mounting media
4. Tweezers, pencil with eraser
5. Filter paper
6. Hot plate
7. Rinzl® brand plastic microscope slides (3 in. × 1 in. × 0.5 mm)
8. A 12-in. length of heavy duty thread
9. One size 5/10 sewing needle
10. Several new Teflon-coated single-edge razor blades

Hands-on Exercise

Hair, fiber, and particulate matter examinations are conducted every day in most working forensic science laboratories across the globe. These samples must be properly prepared before they can be examined with the light microscope. The hair or fiber specimen must be singled out before any procedure can be conducted.

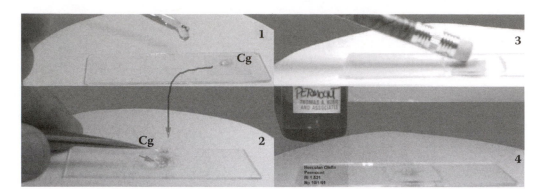

Figure 10.1
Preparation of a wet mount in Permount or Cargille® oil.

Preparation of Wet Mount in Permount® (Figure 10.1)

1. Each student should single out a specimen to be prepared for examination.
2. Place a clean microscope slide on a piece of clean filter paper.
3. Place several drops of Permount on the microscope slide.
4. Place the specimen to be mounted on top of the resin.
5. Place a cover glass with a drop of Permount on its underside onto the specimen.
6. Gently press down with a pencil eraser to remove all the bubbles.
7. Label the preparation and observe the morphology at 100×.

Preparation of Scale Cast Impression in Meltmount® (Figure 10.2)

1. Each student should single out a specimen to be prepared for examination.
2. Adjust a hot plate to 60–70°C (the melting range of Meltmount 1.539).

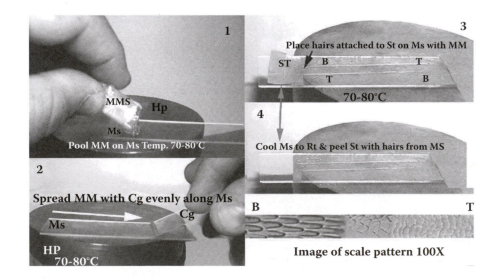

Figure 10.2
A method for the preparation of scale casts in Meltmount.

3. Place a clean microscope slide on the hot plate and gently warm.

4. Apply a Meltmount stick to the slide's surface.

5. Spread the liquefied resin evenly across the slide using another slide.

6. Remove the slide from the hot plate and allow it to cool for 30 seconds.

7. Place the hair to be cast on top of the Meltmount and reheat on the hot plate until the resin liquefies — about 5 sec. Be sure the entire length of the hair is in the resin.

8. Remove the slide from the hot plate and allow the slide to cool completely.

9. Gently peel the hair from the slide. An impression of the hair will remain in the resin.

10. Observe the scale impression with the microscope at 100×.

Preparation of Wet Mount in Meltmount (Figure 10.3)

1. Each student should single out a hair, fiber, or particle to be mounted in Meltmount®; the hair that was just cast in Meltmount (Figure 10.2) is a good specimen to mount.

2. Place a clean microscope slide on the hot plate and gently warm.

3. Apply a Meltmount stick to the slide's surface.

4. Spread the liquefied resin evenly across the microscope slide with another slide.

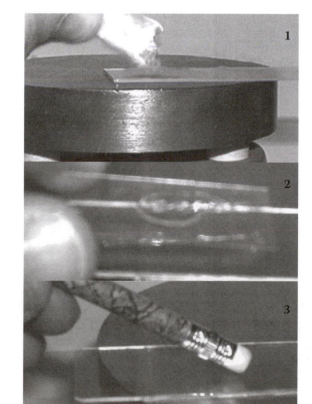

Figure 10.3
Mount hair/fiber in Meltmount.

5. Place a drop of warmed resin on the underside of the cover glass.

6. Place the specimen to be mounted on top of the warm resin.

7. Softly place the cover glass on top of the preparation. Lightly press down on the cover glass with a pencil eraser to remove all the air bubbles.

8. Label the preparation and observe the scale impression with the microscope at 100× (see Figure 10.3).

Preparation of Hair or Fiber Cross Sections on Plastic Slides (Figure 10.4)

1. Each student should single out a tuft of hair or fibers.

2. The needle is double threaded with a length of heavy thread.

3. The threaded needle is used to puncture a 1.0-mm hole in the center of a plastic slide.

4. The sewing needle is then drawn completely through the plastic slide leaving a loop of thread on the opposite side of the clear slide.

5. The needle is then removed from the thread.

6. The hairs or fibers to be cross-sectioned are placed into the loop of thread and the thread is pulled downward until the material is drawn halfway through the slide.

7. An equal amount of the material is left on each side of the plastic microscope slide.

8. A new Teflon-coated single-edge razor blade is used to cut the hair or fiber bundle. The cutting edge of the blade is held at approximately a 30-degree angle to the surface of the slide while being drawn across its surface.

9. The tuft of fibrous material is cut flush to the slide on both sides.

10. The cross sections can be viewed directly or after mounting in an appropriate mounting medium.

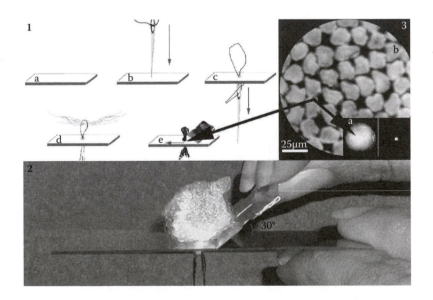

Figure 10.4
The cross-sectioning of fibrous materials with plastic microscope slides.

Name _____ Date _____

Report

1. Write the procedures for the different types of microscope preparations discussed in this experiment.

2. Describe the appearance of all the specimens you prepared for the laboratory.

3. Save all of the preparations by taping them into your laboratory manual or notebook.

TAPE SLIDES TO THIS PAGE.

Experiment 13

Measurement with the Microscope

Teaching Goals

1. The student will learn the procedures and techniques employed by forensic microscopists for making measurements of minute structures with the light microscope.
2. The student will gain experience in manipulating and mounting various types of fibrous and particulate specimens into different mounting media.
3. The student will gain additional experience in utilizing the light microscope for identification purposes.

Background Knowledge

Minuscule lengths of synthetic and natural fibers, human and animal hairs, grains of minerals, fragments of glass, wafers of smokeless powder, red blood cells, skin cells, and wood fibers are just a few categories of the trace evidence materials that are examined, identified, and compared by forensic microscopists in their daily work.

The physical morphology and the size of the structures comprising many of these types of trace materials are important features used in the identification and comparison of questioned and known specimens. It is often necessary to measure the thickness, diameter, length, width, and angles of the tiny structures comprising many of these materials.

The following exercise elucidates the method used to measure the length, width, and diameter of the tiny structures making up the various types of trace evidence.

Equipment and Supplies

1. Questioned specimens collected in Experiment 9
2. Ocular and stage micrometers*
3. Glass microscope slides and cover glasses
4. Assorted mounting media
5. Known synthetic fiber standards
6. Fine tweezers, filter paper, and a pencil with an eraser
7. Light microscope

---

* Your microscope may have the ocular micrometer already installed. Follow the instructions of your professor.

Hands-on Exercise

1. Measurement of small linear distances, angles, and areas with the microscope is known as micrometry. Quantitative measurement with the microscope involves the use of various types of ocular scales, some of which are calibrated with a stage micrometer. The normal unit of measurement for length, width, and thickness is the micrometer (μm). One micrometer is equal to one-millionth of a meter or 1×10^{-6} m.

2. There are two distinct types of micrometers used with a microscope. The first is the ocular micrometer, which is usually an arbitrary scale produced on a round disc of glass that is placed in the primary focal plane of the ocular. The ocular scale can take on many configurations. It can be a crosshair, a small ruler made from a vertical or horizontal line divided into 100 equal divisions, a crosshair made by combining a vertical and horizontal ruler, a grid divided into 100 equal squares, and many other forms. The second type of micrometer is the stage micrometer. This is typically a scale of known length, normally 1 mm, divided into 100 equal divisions, with each division's value equal to 10 μm (see Figure 13.1).

3. The ocular micrometer is positioned in the eyepiece as shown in Figure 13.2.

4. In order to determine the value for each ocular micrometer unit (OMU) you must align the image of the stage micrometer with that of the ocular micrometer, as shown in Figure 13.3. Both micrometers are aligned as a vernier and the number of OMUs and stage micrometer units (SMUs) are counted and recorded. The value (in micrometers) for one OMU is equal to the number of SMUs times the value for each SMU divided by the number of OMUs.

5. The previous procedure must be carried out for each objective lens. The value for each objective should be recorded and taped on the base of your microscope for quick reference. This value will remain the same as long as the objectives, oculars, and body tube or head are not changed or altered. Figure 13.4 shows an actual image taken through the microscope when a 20× objective was being calibrated (see Table 13.1).

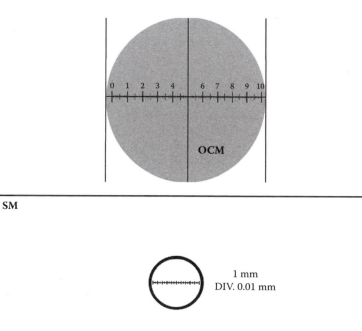

Figure 13.1
The two common types of micrometer scales: the ocular micrometer and the stage micrometer.

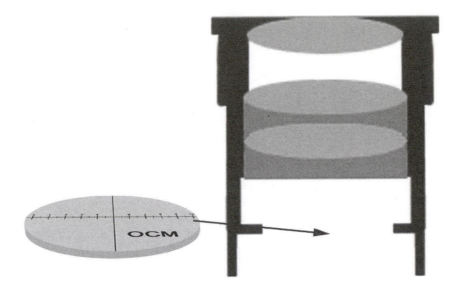

Figure 13.2
Placement of the ocular micrometer into the eyepiece or ocular.

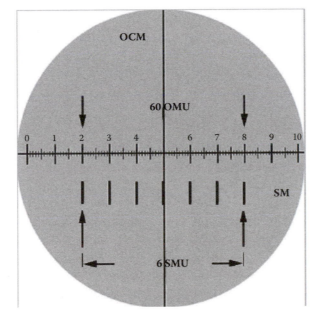

Figure 13.3
Typical ocular and stage micrometers. The calibration of an ocular micrometer is shown. In order to determine the value of each OMU, one must align the image of the stage micrometer with that of the ocular micrometer. Both micrometers are aligned and the number of OMUs and SMUs are counted and recorded. The value (in micrometers) for one OMU is equal to the number of SMUs times the value for each SMU divided by the number of OMUs.

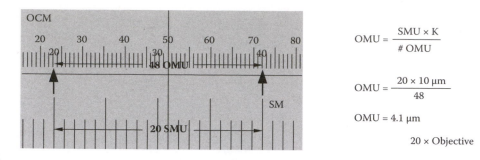

$$OMU = \frac{SMU \times K}{\# \, OMU}$$

$$OMU = \frac{20 \times 10 \, \mu m}{48}$$

$$OMU = 4.1 \, \mu m$$

20 × Objective

Figure 13.4

To determine the value of one OMU for a 20× objective, first the objective must be focused on a stage micrometer. Next, the images of the ocular micrometer and stage micrometer are aligned as a vernier. The number of OMUs and SMUs that align with each other are counted and recorded. The value (in micrometers) for one OMU is equal to the number of SMUs (20) times the value of each SMU (10 μm) divided by the number of OMUs (48). In this case, 1 OMU = 4.1 μm. Repeat the calibration procedure for each microscope objective.

TABLE 13.1
Calibration Table for Each Objective on the Microscope

Objectives lens	No. of OMU	No. of SMU	K = constant	Value of 1 OMU
4×				
10×				
20×	48	20	10 μm	4.1 μm
40×				
100×				

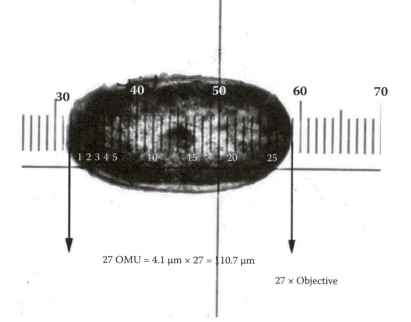

27 OMU = 4.1 μm × 27 = 110.7 μm

27 × Objective

Figure 13.5

The measuring of a human head hair cross section with the calibrated ocular micrometer and 20× objective shown in Figure 13.4. To determine the hair's width, the image of the hair cross section is aligned with the image of the ocular micrometer. The number of OMUs that align with the hair's cross section are counted and recorded. The value (in micrometers) for the hair's cross section is equal to the number of OMUs (27) times the value for each μm (4.1 μm). In this case, the hair's cross section is equal to 27 × 4.1 μm, or 110.7 μm.

Name _____ Date _____

Report

1. All students must document in their notebooks the correct procedures for calibration of an ocular micrometer with a stage micrometer.

2. Calibrate all of the objectives available on your microscope.

3. Prepare Table 13.2, listing the values of one OMU for each objective lens calibrated.

TABLE 13.2
Calibration Table for Each Objective on Your Microscope

Objectives lens	No. of OMU	No. of SMU	K = constant	Value of 1 OMU
4×				
10×				
20×				
40×				
100×				

4. Each student must measure the thickness of five specimens. One hair, two fibers, and two other types of specimens of your own choice (e.g., pollen grain, mineral grain, yeast mold, plant hair). Each measurement must be clearly and completely documented in your notebook.

5. Compare and contrast the results from each student or group. Calculate the mean, medium, and mode for each type of specimen.

Experiment 

Examination of Trace Quantities of Synthetic Fibers

Teaching Goals

The goals of this exercise are to introduce the student to the procedures used in the preparation, mounting, examination, and characterization of synthetic fibers. In addition, the student will gain additional experience with the light microscope.

Background Knowledge

Dr. Edmond Locard, a famous 20th-century French forensic scientist, developed the "principle of mutual exchange." In his principle, Locard proposes that whenever two people or a person and an object or place come into contact, there is a mutual exchange of trace amounts of material. Locard believed that every crime could be solved by studying the trace evidential materials transferred during the event.

Synthetic fibers are often encountered in forensic casework. Synthetic fibers used in the manufacture of wearing apparel, office and household draperies, textiles, and carpets are easily transferred between people, places, and things. Today, forensic scientists apply Locard's principle to the study of microscopic traces of synthetic fibers to help associate the people, places, and things involved in a crime, to reconstruct the events of a crime, and to solve the crime. This exercise will introduce you to the methods used by forensic microscopists to examine synthetic fibers.

Equipment and Supplies

1. Questioned fiber specimens collected in Experiment 9
2. Glass microscope slides and cover glasses
3. Mounting media
4. Known synthetic fiber standards
5. Fine tweezers and a pencil with an eraser
6. Filter paper
7. Light microscope

Hands-on Exercise

One must clearly see a specimen in order to collect fundamental information about its appearance or physical structure. The extent to which a colorless to lightly colored fiber can be seen when immersed in a colorless or nearly colorless mounting medium is known as relief. If the specimen and the mounting medium

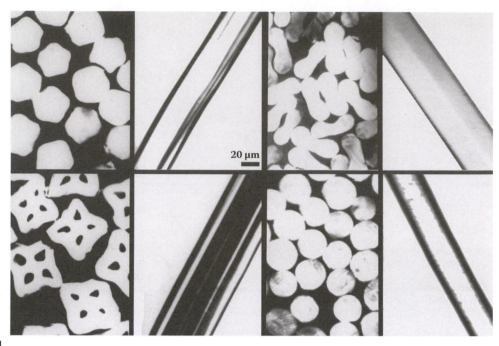

Figure 14.1

A few of the fiber cross-sectional shapes available today. Note their cross-sectional shape and the corresponding longitudinal appearance.

have the same refractive index (RI), the specimen will not be visible when viewed with a light microscope under plane polarized light. If the RI of the mounting medium is close to that of the specimen, a low-relief condition results. If the RI is somewhat different from that of the specimen, moderate relief occurs and the specimen will be fairly visible in the mounting medium. If the difference between the specimen and mounting medium refractive indices are large, high relief will result and the specimen will appear to stand out.

1. Use fine tweezers to remove the suspected synthetic fibers collected in Experiment 9 from the container in which they are stored. Observe them at low magnification under a stereomicroscope or magnifying glass.

2. Prepare a wet mount of the questioned fiber specimens collected in Experiment 9 by following the procedure outlined in Experiment 10. The mount should be prepared in a Cargille oil having an RI of approximately 1.540 at 25°C for the sodium D line (589 nm).

3. Enter fiber morphology observations on the Fiber Data Sheet (Appendix A).

    a. Note the longitudinal appearance. Are the fibers smooth, serrated, round, square, pentagonal, lobed, or irregular (see Figure 14.1)?

    b. If possible, determine the cross-sectional shape of the fibers by their longitudinal appearance.

    c. Measure the diameter thickness of the fiber (in micrometers) as instructed in Experiment 13.

4. Enter all fiber optical data on the Fiber Data Sheet (Appendix A). To determine whether a fiber is anisotropic (has more than one refractive index), place a polarizing light filter under the microscope condenser with its preferred vibration direction in the east–west position (see Figure 14.2). As the fiber is observed at 100×, slowly rotate the filter until its preferred direction is in the north–south position. If the contrast between the fiber and the mounting appears to change, as in Figure 14.2, the fiber is birefringent (has two or more RIs). If no change in the fiber appearance is noted, the fiber is isotropic (has one RI).

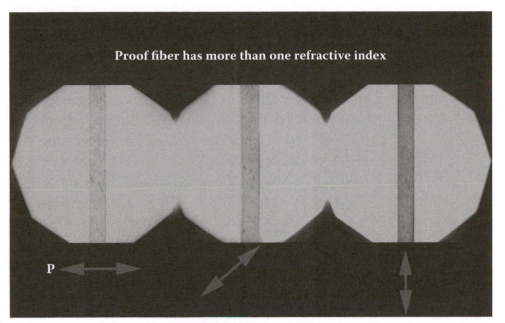

Figure 14.2

A fiber displaying (left) low relief and (right) high relief. To determine if the fiber is anisotropic (has more than one RI), a polarizing light filter is placed under the microscope's condenser with its preferred vibration direction in the east–west position (left). While the fiber is being observed at 100×, the filter is slowly rotated until its preferred direction is in the north–south position. If the contrast between the fiber and the mounting medium changes, as in the above photomicrograph, the fiber is birefringent (has two or more RIs). If there is no change in the fibers appearance, the fiber is isotropic (has one RI).

    a. Is the fiber easily visible in the mounting medium? What is its degree of relief? High-relief fiber is distinctly visible; a low relief fiber image blends into the mounting medium.

    b. Does the fiber have more than one RI (anisotropic) (see Figure 14.2) or only one RI (isotropic)?

    c. If the fiber is anisotropic, determine its relative RIs in the mounting medium by the Becke line method (see Figure 14.3).

    d. To determine the generic classification of an unknown fiber, the collected information on the Fiber Data Sheet is compared to the data in Table 14.1 and the flow chart (Figure 14.4).

    e. Identify each class of fiber specimen in the same manner. If a comparison of the fibers is desired, the questioned and known specimens can be compared side by side on a comparison microscope.

5. Each student should prepare a wet mount of the questioned fiber specimen collected in Experiment 9 by following the procedure outlined in Experiment 10. The mount should be prepared in a Cargille oil having an RI of approximately 1.540 at 25°C for the sodium D line (589 nm).

6. Fiber morphology: enter all observations on the Fiber Data Sheet and in your notebook.

7. Fiber optical data: enter all observations on the Fiber Data Sheet and in your notebook.

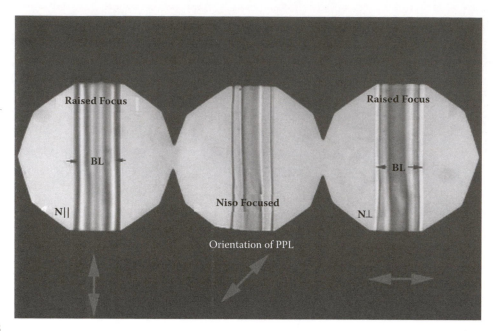

Figure 14.3
Determination of the relative refractive index (RRI) for a fiber specimen's N | | (left) and N⊥ (right) directions. The procedure involves observation of the Becke line movement in plane polarized light for both the N | | and N orientations of the fiber. After focusing the specimen (center), the orientation of the polarizing filter is made parallel to the fiber's long direction (N | |). Next, the microscope's focus is raised and the movement of the Becke line is noted (right). Then the orientation of the polarizing filter is adjusted until its vibration direction is perpendicular to that of the fibers width (N| |). The microscope's focus is raised and the movement of the Becke line is noted in this orientation (left). In this case, the fiber's N | | is greater than the mounting medium and the fiber's N⊥ is less than the mounting medium.

TABLE 14.1

A Few Generic Classes of Synthetic Fibers Commonly Seen in Forensic Casework

Generic Class	N ||/N⊥ Relative Refractive Index	Birefringence	Relief	Cross section
Acrylic	Both < 1.539	Very low	Low–medium	Bean, dog bone, mushroom, round
Polyamide (nylon 6, 6.6)	N|| > 1.539, N||< 1.539	High	Low–medium	Round, trilobal, tetralobal
Glass, mineral wool	Isotropic	None	Low	Round, off round, irregular
Olefin (propylene)	Range 1.510–1.620, both < 1.539	Medium	Low–medium	Round, trilobal, delta, flat
Polyester	N || > 1.539, N⊥ close to 1.539	Very high	Low–high	Round, ovoid, polygonal, donut, trilobal, swollen ribbon

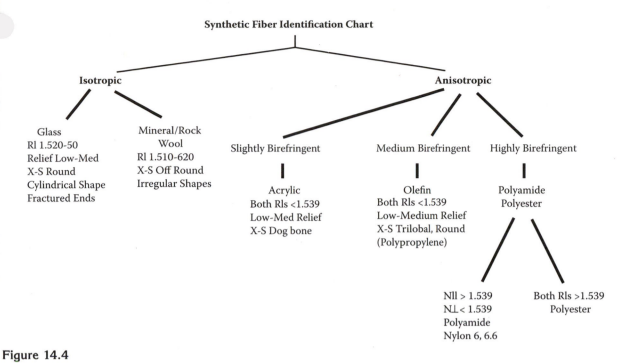

Figure 14.4
Synthetic fiber flow chart.

Name _____ Date _____

Report

1. All students must document in their notebooks the correct procedures for the microscopic examination of synthetic fibers.

2. Make a sketch of your fiber specimen, including both longitudinal and cross-sectional views.

3. Make a list of the characteristics used to identify each class of synthetic fiber.

APPENDIX A

Fiber Data Sheet

Fiber Morphology (write in and/or circle)

Longitudinal morphology: smooth ____ striated ____ irregular ____ other _____

Cross-sectional shape _____

Diameter of lobe(s) (in μm): _____

Reflected color _____

Transparency _____

Transmitted color _____

Length (in mm) _____

Surface texture _____

Staple length or continuous _____

Type of coloration: pigment or organic dye _____

Treatments: twisted, crimped, melted _____

Presence, extent, and type of dulling agent _____

Other manufactured artifacts (i.e., gas bubbles, fish eye) _____

Optical Data

Relative refractive index (relative to medium 1.539-40 or another mounting medium with known RI):

(1) N parallel (N‖)   above   below   near   equal to _____

(2) N perpendicular (N⊥)   above   below   near   equal to _____

With polarizer: Is fiber isotropic or anisotropic? _____

Interference colors: _____

Fluorescence: UV _____

Other information _____

Experiment

Basics of Photography

Teaching Goals

Students will be introduced to the basics of traditional photography utilizing a 35 mm camera. Students will learn the basic techniques of photography and the operation of a 35 mm camera. Students will also gain additional knowledge and practice in documenting physical evidence.

Background Knowledge

Documentation of an unaltered crime scene and its physical evidence is a crucial part of any investigation. Each scene and its physical evidence must be photographed in their entirety. As each item of evidence is discovered, it must be photographed to show its position and location at the crime scene as well as its relationship to the overall scene. Therefore it is extremely important that a forensic scientist be proficient in the basic methods of photography. This lesson and Experiment 16 are designed to give you this rudimentary knowledge.

Equipment and Supplies

1. Camera (35 mm) with assorted lenses
2. Tripod or photographic stand
3. Different speeds (ISO) of 35 mm film
4. Different types of evidence displays
5. Scales

Hands-on Exercise

1. Study the parts of a 35 mm camera (Figure 15.1) and the glossary of terms.
2. Practice operating the camera as directed by your instructor.
3. Photograph the evidence displays using different lighting techniques.
4. Develop the film as instructed. If black-and-white film is used, refer to Experiment 16.
5. Some important concepts to consider:
   a. How a lens forms an image (Figure 15.2).
   b. How a camera works (Figure 15.3).
   c. Film speed (ISO).
   d. Film image size.

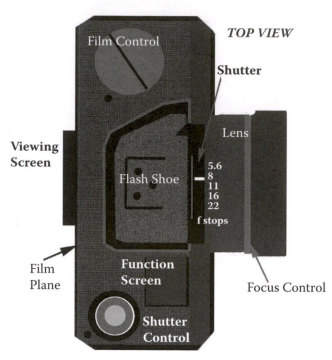

Figure 15.1
Top view of a 35 mm camera.

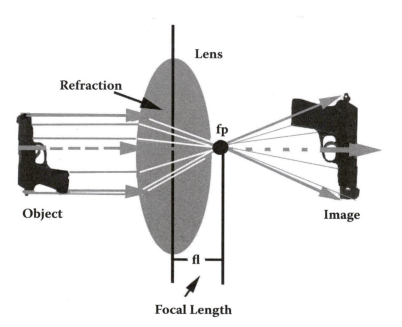

Figure 15.2
Formation of an image by a lens. An image is formed when rays of light traveling from an object enter a lens and are refracted or bent. The converging rays of bent light form an inverted real image of the object. The image of the object is enlarged if it is formed at a distance greater than two focal lengths from the lens.

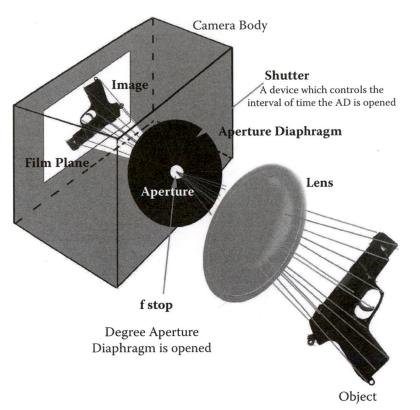

Figure 15.3
The basic elements of a 35 mm camera.

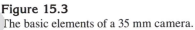

e. Focusing the image onto the film plane.

f. Coverage of the object area.

g. Size of the aperture opening or f-stop.

h. Maximum depth of field obtainable.

i. Shutter speed or exposure time.

j. True size of object in print (scale).

k. Relationship between exposure time and f-stop.

l. Working distance from the object to the front of the lens.

Name _____ Date _____

Report

1. Define refraction.

2. How does the camera form an image?

3. What is the primary difference in speed between ISO 50 and ISO 200 film, i.e., how much less or more light does the ISO 200 film require than ISO 50?

4. Why is oblique illumination so useful in forensic photography?

5. I used a film with an ISO value of 50 to photograph an object while my camera was set to an f-stop of 5.6 and an exposure time of 1 second. The resulting image of the object was somewhat blurry, yet my camera was sharply focused. How can I improve the image?

6. Develop your photographs based on the techniques in Experiment 16 and use them to prepare a display.

Photography Glossary

Aberration — An optical defect in a lens.

Acetate — Cellulose acetate film base, which supports film emulsion.

Aperture — Lens opening or f-stop, which regulates the amount of light entering a camera.

Bellows — Adjustable tube or chamber that connects the lens to the camera body.

Cable release — A flexible wire capable of activating a shutter when depressed.

Cleaning agent — Solution that reduces the washing time by chemically neutralizing fixing agents.

Contact print — A negative is placed on photographic paper and controlled light passes through the negative, producing an image of the same size on the paper.

Depth of field — The distance between the nearest and farthest objects in sharp focus in a photograph.

Emulsion — A thin gelatin coating containing light-sensitive silver salts which is placed on film or paper. The manufacturer assigns an emulsion value (speed) that is used to determine proper exposure.

Fix — To dissolve undeveloped silver salts on film through the use of hypo.

Focal length — The distance from the lens to the film plane when the lens is focused on an object at infinity.

Focus — Adjusting the lens-to-film distance in order to sharply define a subject.

f-stop — Designation indicating the aperture size (opening) and the amount of light passing through a lens.

Grain — Clumped silver particles that appear as light spots in photographic prints.

Highlights — The brightest area of a subject.

Hypo — A bath made chiefly of sodium thiosulfate, used for fixing.

ISO — Rating of the film emulsion relating to its sensitivity to light.

Light meter — An instrument used to measure light reflected from or falling on a subject.

Macrophotography — Close-up photography making the object appear larger.

Negative — Developed film that shows the image in reverse tones from the original subject.

Print — A positive rendition of a subject, usually made from a negative.

Refraction — Bending of light rays when they pass from one medium to another.

Stop bath — Weak acetic acid solution used to terminate (stop) development of film or paper.

Experiment 16

Black-and-White Film Development

Teaching Goals

The objective of this lesson is to allow the student to learn the basic procedures necessary to develop negatives and prepare black-and-white photographic prints.

Equipment and Supplies

1. Negative developing solutions and equipment for 35 mm film
2. Photographic paper, developing solutions, and trays
3. Darkroom with sinks, running water, safelights, and counter space
4. Enlarger for 35 mm film
5. Thermometer
6. Magnifying glass
7. Ruler or appropriate scale
8. Timer

Hands-on Exercise

When film is exposed to light, the silver compounds embedded in the film emulsion undergo a chemical reaction and produce an image that can be made visible by application of developing chemicals. A negative developer reacts with the emulsion and removes the unaffected silver halides. A stop bath halts the developer action and a fixing compound sets the silver image of the subject on the film. Figure 16.1 shows the equipment necessary to develop black-and-white prints. Figure 16.2 shows the components of a darkroom.

Film Development

1. In the darkroom, open the camera and remove the film exposed during Experiment 15.
2. Wind the 35 mm film onto a film reel and place the reel in a developing canister.
3. After securing the lid on the canister, turn on the room lights.
4. Pour the appropriate developer into the canister. The developing chemicals should be kept at room temperature and premixed according to the manufacturer's instructions. Allow the film to develop for 3 to 5 min.
5. Pour out the developer into a collection container and rinse the film with water.
6. Pour out the water rinse and add stop bath to the canister. Wait 30 sec, pour out the stop bath, and rinse the film with water.
7. Discard the water rinse and add fixing solution to the canister. Fix the film for 10 min.

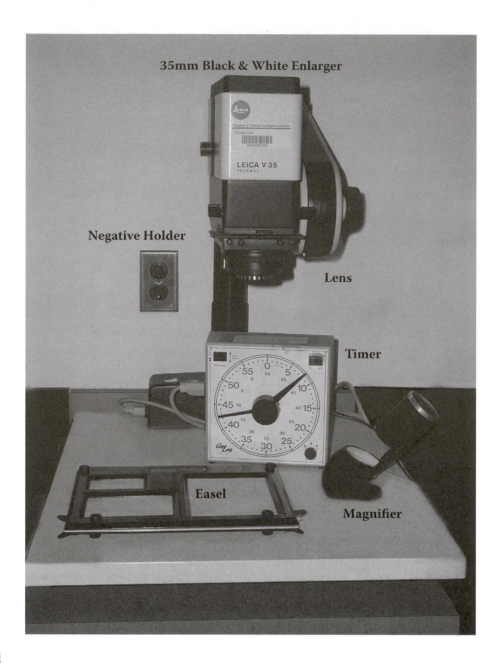

Figure 16.1
Equipment necessary for developing black-and-white photographic prints.

8. Remove the film from the canister and wash with water for 30 min.
9. Squeegee the film surface before hanging the film to dry.

Enlargement of Negatives (Figure 16.1)

1. Place the dried 35 mm negative strip into the negative carrier of the enlarger.
2. Place the negative carrier into the enlarger.
3. The emulsion side of the negative must face downward, toward the enlargement easel.

Figure 16.2
Darkroom photographic printing equipment and solutions: (1) developer tray, (2) stop bath tray, (3) fixing solution tray, (4) washing tray, (5) squeegee table, and (6) drum-type print drier.

4. Turn off the room lights; work only with safelights.
5. Place a sheet of old print paper into an easel board to aid in viewing, focusing, and sizing of the image.
6. Turn the enlarger on and open the lens to its widest aperture.
7. To produce the desired photographic print, move the enlarger unit up or down to achieve the proper size image. Use a scale to size the image.
8. Move the bellows up or down to properly focus the image.
9. Set the aperture stop to f8 and the exposure time to 1 sec.
10. Turn the enlarger off and replace the old print paper in the easel with a sheet of unexposed photographic paper.
11. Expose the paper and develop according to the procedure below.

Print Development (Figure 16.2)

1. Place the exposed paper in the developer. The developing chemicals should be premixed according to the manufacturer's instructions and at room temperature. The image should form in 1 to 2 min.
2. Wash the print and place it in the stop bath for 5 sec.
3. Wash the print and place it in the fix for 10 min.
4. Remove and wash the print in running water for 30 min.
5. Squeegee, drain, and dry with a photographic drying drum.
6. Make aperture, exposure, and filter adjustments to the enlarger until the optimum results are obtained.

Name _____ Date _____

Report

1. Record all observations in your notebook.

2. Prepare a court display of your evidence photographs as instructed by your professor.

3. Hand in your court display at the beginning of the next laboratory session or as otherwise instructed by your professor.

Collection of Footwear Evidence

Teaching Goals

The purpose of this experiment is to allow the student to learn the basic procedures necessary for the recovery of footwear evidence from crime scenes. The student will also gain knowledge of the jargon of footwear examination. In addition, the student will gain added familiarity with the concepts of identity and individualization.

Background Knowledge

Footwear evidence can be defined as material surfaces that have retained an image of footwear under sole and heel patterns as well as wear patterns and individualizing characteristics. These images are acquired by means of residue transfer from the sole of a piece of footwear to a surface or by the compression of loose material by the sole of a piece of footwear. Therefore footwear evidence can appear as either two- or three-dimensional prints.

The primary purpose for conducting a footwear examination is to associate people with a crime scene. Some additional objectives for conducting footwear evaluations in the laboratory may include the following:

- Examination of the questioned impression to supply investigative information (i.e., type of footwear, size of the shoe that made a print, manufacturer, model, general type).
- Comparison of the questioned impressions with known footwear impressions to include or exclude suspects.
- Reconstruction of the crime.

In this lesson, students will cast a three-dimensional footwear imprint in soil. In Experiment 18, the identification and comparison of footwear evidence will be discussed.

Equipment and Supplies

1. Questioned footwear residue prints (two-dimensional) and questioned footwear impressions (three-dimensional) provided by the instructor

2. Supplies necessary for the documentation of footwear evidence:
   - 35 mm camera
   - Film
   - Photographic accessories (i.e., tripod, level)
   - Lighting or illumination sources (ultraviolet, infrared, visible monochromatic)

3. Supplies necessary for the collection of two- and three-dimensional footwear evidence:

- Clear lifters (11 × 14 in. and 8 × 10 in.)
- Gel lifters (black, white, and clear)
- Electrostatic dust lifting device kit — includes roller, lifting foil, alligator clips, ruler, charging unit, metal tray, instructions
- Casting materials (dental stone, silicone)
- Plastic Ziploc bags
- Plastic bottles filled with water

4. Magnifying glass
5. Various light sources

Hands-on Exercise

1. Locate the questioned footwear print. If you are at a crime scene, employ the appropriate searching procedures to locate the print.
2. If a two-dimensional residue print is the goal, use the oblique lighting technique to locate latent footwear prints. Figure 17.1 shows a photographic setup for revealing a latent footwear print. The setup consists of oblique

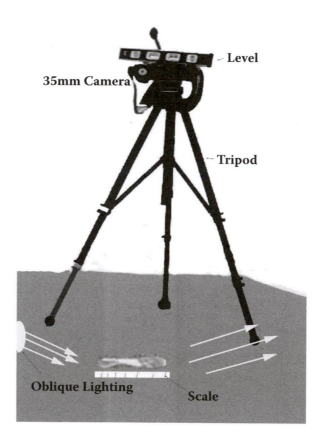

Figure 17.1
The photographic setup for documenting a latent residue footwear print using oblique lighting, a level, and a 35 mm camera on a tripod with high-contrast black-and-white film. A digital camera can be used. Remember to fill the frame with the image of the print and always use a scale.

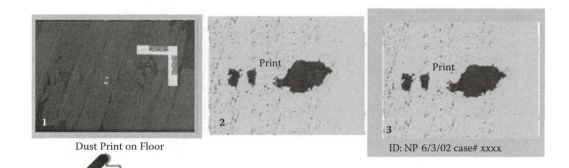

Dust Print on Floor

ID: NP 6/3/02 case# xxxx

1. Place clear lifter over print and press down with roller.
2. Remove clear print lifter.
3. Place lift on a sheet of contrasting glossy paper.

Figure 17.2
The procedure for lifting a residue dust print with a clear plastic lifter.

lighting, a level, a 35 mm camera on a tripod, and high-contrast black-and-white film. A digital camera can be used. Fill the frame with the image and always use a scale.

3. Document the footwear print by photographing it.

4. If you are working with a dust print, first document it photographically, then lift it with a clear plastic adhesive lifter. Figure 17.2 shows the procedure.

5. If available and appropriate, use an electrostatic dust lifting device to lift the dust footwear print (Figure 17.3). The procedure is as follows:

   a. If the print is on a movable object, such as a newspaper, place the object on a metal tray. If the print is not on a movable object, electrically ground the area immediately around the print.

   b. Place the dark side of the foil on top of the object; apply the electrodes.

   c. Apply an electrical charge to the foil by engaging the power supply switch.

   d. Press down with a roller; move the roller back and forth until the foil is flat.

   e. Turn the foil over to reveal the print.

   f. Review the instructions for the electrostatic dust lifting device.

6. If a three-dimensional impression is found, use the oblique lighting technique to photograph the impression.

7. Cast the impression using the technique shown in Figure 17.4. The general rules are as follows:

   a. Shallow impressions in sand or soil less than 1 in. deep should be cast with dental stone or silicone (Figure 17.5). Dental stone is a specially formulated casting plaster. It can be mixed in a plastic bag (2 lb of dental stone and 12 oz water) and does not require reinforcement.

   b. Deep impressions should be cast with dental stone (a specially formulated casting plaster). Dental stone does not require any reinforcement as does plaster. Dental stone is mixed in a plastic bag: 2 lb DS to 12 oz water.

   c. Impressions in snow and ice can be cast with paraffin or sulfur.

   d. The method for casting footwear impressions in soil. After photographing, remove loose objects and spray the print with clear lacquer or snow wax (if impression is in snow) (see Figure 17.4). Figure 17.5 shows a finished dental stone cast; note the fine details.

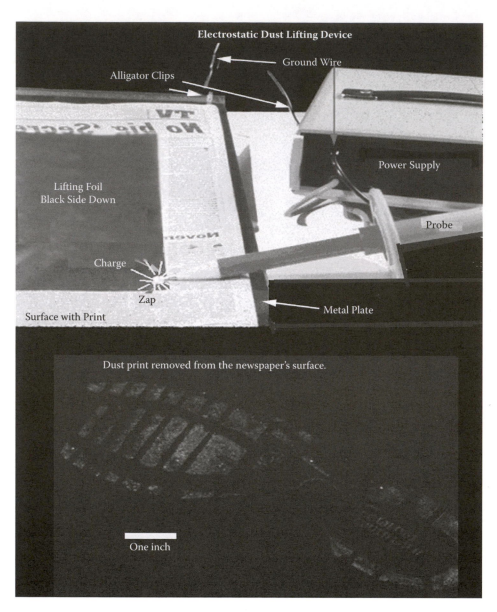

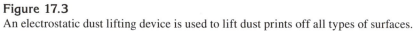

Figure 17.3
An electrostatic dust lifting device is used to lift dust prints off all types of surfaces.

1 Impression in Soil

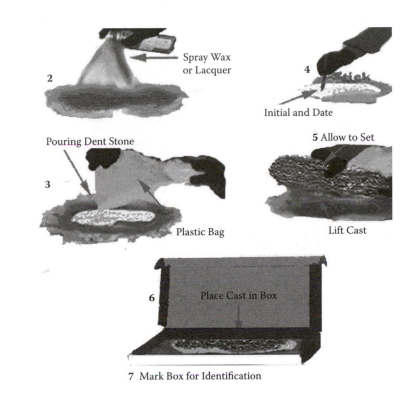

2 Spray Wax or Lacquer

4 Initial and Date

Pouring Dent Stone

3 Plastic Bag

5 Allow to Set

Lift Cast

6 Place Cast in Box

7 Mark Box for Identification

Figure 17.4
Casting a three-dimensional footwear print.

Figure 17.5
Finished dental stone cast of a footwear impression.

Name _____ Date _____

Report

1. Record all observations in your notebook.

2. Draw a rough sketch in your notebook.

3. For homework, using the data collected and documented in your rough sketch, your lifts or casts, your notes, and your photographs, prepare a detailed scale drawing with direction (north), a key, and any necessary elements (refer to Experiment 8).

4. Hand in your drawing with all exemplars at the beginning of the next laboratory session or as otherwise instructed by your professor.

Identification and Comparison of Footwear Impressions

Teaching Goals

The primary objective of this lesson is to allow the student to learn the concepts, rationale, and procedures necessary to identify and compare footwear evidence. In addition, each student will gain experience in making court displays.

Background Knowledge

Footwear evidence is frequently encountered at crime scenes. Footwear impression evidence can be very useful information in forensic investigations. Footwear print evidence can be used by forensic investigators to:

- Lead to the identity of a suspect
- Eliminate a suspect
- Help determine the brand of shoe with the use of a database
- Help determine a general or specific shoe size
- Positively identify the shoe (unique characteristics)
- Help prove the presence of a suspect at the scene of a crime
- Help reconstruct a crime scene
- Show the number of suspects involved
- Show involvement in the crime
- Show the time frame in which the impression was made (wet to dry mud)
- Show the sequence of events
- Help verify or disprove alibis

Hands-on Exercise

Preparation of Known Standards

1. Moisten the outsole of a known item of footwear with water. Figure 18.1 shows known and standard footwear.
2. Tap off excess water; remove remaining water with a dry cloth.
3. Apply fingerprint powder to the outsole with a fiberglass brush. Three or more applications may be required.
4. Place a piece of clear adhesive lifter or a moistened piece of roll transport film, adhesive side up, onto a cushioning pad.
5. Put a protective plastic bag on your foot and place your foot into the known footwear.

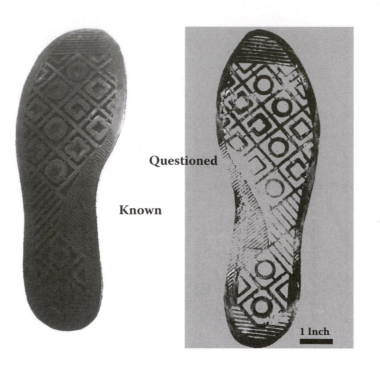

Figure 18.1
Known piece of footwear and standard.

6. Attempt to walk normally and step onto the film.

7. Place the film on a table and allow it to air dry. A blow-dryer can be used to accelerate drying. If adhesive lifter is used, cover it with a clear plastic sheet and press down with a roller (see Experiment 17).

Questioned Print Examination

Typical features to be observed in a footwear examination include class characteristics produced during manufacturing (design of the sole), wear patterns produced during normal wear (e.g., worn-down heel areas), and accidental or individualizing marks or patterns produced by random chance during normal wear. Figure 18.2 shows class, wear, individualizing characteristics, and patterns.

Accidental or identifying characteristics are cuts, tears, gouges, adherence of foreign bodies (gum, pebbles, fragments of glass), and wear marks that randomly appear when a shoe is worn. These characteristics can often individualize a known shoe to a questioned print.

Class characteristics are intentional or unavoidable characteristics that repeat during the manufacturing process and appear on all shoes. The characteristics are size, shape, pattern design, and manufacturing characteristics.

Wear patterns are the result of random removal of material from the outer sole and heel as shoes are worn. Wear patterns serve as supplemental identifying features. Most people, due to their body's structure, physical condition, health, and the manner in which they walk, usually wear out their shoes in a consistent manner. As a result, the wear patterns that form on an individual's shoes can help associate questioned footwear prints to known pieces of footwear.

Important Concepts

1. Class characteristics are intentional or unavoidable characteristics that repeat during the manufacturing process and are shared by two or more shoes. These are size, shape, pattern design, and manufacturing characteristics.

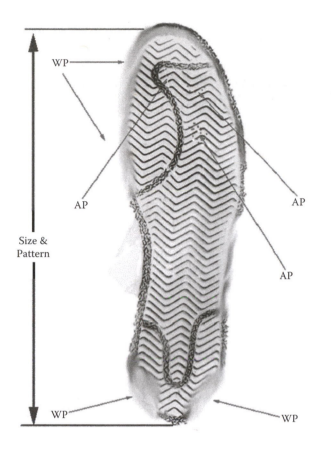

Figure 18.2
The class, wear, and individualizing characteristics and patterns one should examine during a footwear examination.

2. Wear patterns are the result of the random removal of material from the outer sole and heel as shoes are worn. Wear patterns are good supplemental identifying factors. Most people usually wear out their shoes in a consistent manner. Consequently, wear patterns can be used to help associate questioned footwear prints to known pieces of footwear.

3. Accidental or identifying characteristics include cuts, tears, gouges, foreign bodies (gum, pebbles, fragments of glass), and wear marks that occur randomly as a shoe is being worn. These are often used to individualize a known shoe to a questioned print.

Evaluation Process

1. The questioned impression is always compared to the known (not the other way around).

2. Compare the class, wear, and accidental features of the questioned to the known.

3. Since identifying characteristics are constantly being lost and formed with use of the footwear, it is possible, and likely, that characteristics will not be found in both the questioned and known impressions. Depending upon the depth and location of the characteristic, it may be eliminated in 1 day to 2 weeks of continuous wear. In addition, characteristics found may have been altered in shape and size.

Conclusions

- For a positive identification, all class characteristics of the questioned impression must be seen in the known impression; one or more identifying characteristics must appear on both the questioned and known impressions.
- In a positive elimination, known footwear will not possess the same class features as questioned prints.
- Not suitable for comparison means that not enough of the questioned print is available for examination.

Name _____ Date _____

Report

1. Report your findings in your notebook.

2. Justify your conclusions.

3. Make a list of matching features between the questioned and known prints.

4. Prepare court displays. Figure 18.3 shows a residue print prepared for use in court. Figure 18.4 shows a cla
match comparison.

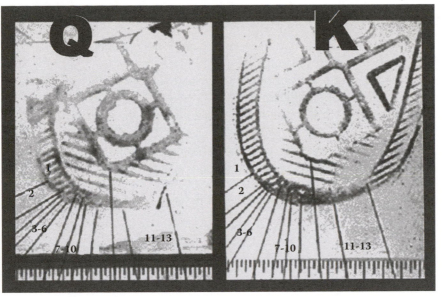

Figure 18.3
Display of a residue print prepared for court testimony.

Figure 18.4
Court display of a cast impression showing a class match.

Footwear Glossary

Heel — A separate component attached to the rear portion of the outsole.

Heel area — The rear portion of the outsole

Logo — A name, design, or pattern that often appears on the sides or the outsole of athletic shoes; it is usually a trademark of the manufacturer.

Midsole — The component on some shoes that is placed between the outsole and the shoe upper.

Outsole — The extreme bottom layer of the shoe; the layer that makes contact with the ground.

Shoe upper — The components of the shoe excluding the outsole and the midsole.

Toe bumper guard — A thick strip of rubber placed across the top of the toe box or the vamp to increase the strength and durability of the shoe in that area.

Tongue — A strip of material that covers the instep of the foot and lies beneath the shoelaces.

Experiment

Tool Mark Examination

Teaching Goals

The objective of this lesson is to provide the student with the concepts, rationale, and procedures necessary to recognize, document, collect, and compare tool mark evidence.

Background Knowledge

Tool marks can be defined as any impression, gouge, gash, slash, groove, channel, dent, indentation, hole, scrape, graze, abrasion, scratch, cut, or striation that has been made on any object, person, or thing by the application of a tool. It is important to note that a tool can be defined as any device or implement that is capable of producing the questioned marks.

In forensic laboratories, firearms and tool mark examiners use a reflective light comparison microscope to compare questioned tool marks found at crime scenes with known tool marks made by suspected tools. Figure 19.1 shows a firearms/tool mark comparison microscope.

Equipment and Supplies

1. Questioned tool marks as provided by your instructor
2. Questioned tools as provided by the instructor
3. Photographic equipment
4. Magnifying glass and stereomicroscope
5. Materials for making standard (known) tool marks (i.e., aluminum sheets 1 mm thick)
6. Casting silicone
7. Scale or ruler

Hands-on Exercise

Questioned Tool Mark Examination

1. Examine the item you receive for any obvious tool marks and any foreign or unusual marks.
2. Document and photograph the questioned marks (Figure 19.2).
3. Examine the questioned marks *in situ* using oblique light.
4. Prepare the casting silicone as per the manufacturer's instructions (Figure 19.3).
5. Apply the casting silicone as shown in Figure 19.4.

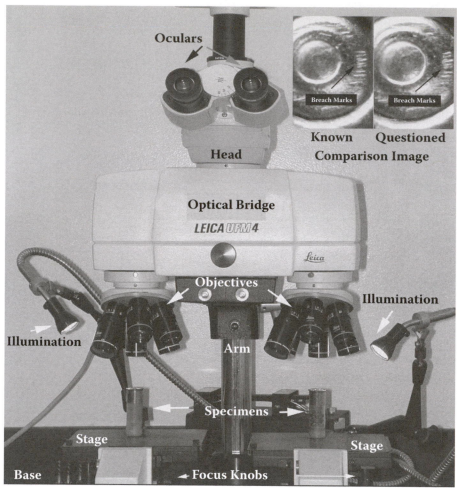

Figure 19.1
A firearms/tool mark comparison microscope.

6. Allow the silicone to set, about 15 minutes, and photograph.

7. Remove the cast for examination and comparison with known tools (Figure 19.5).

Preliminary Examination of Questioned Tool Marks

1. If possible, establish the questioned tool marks' class characteristics (i.e., pattern, size, length, type of tool).

2. Determine the questioned print's wear areas and overall wear pattern.

3. Examine the questioned print's individualizing and accidental characteristics.

Preparation of Known Standards

1. Remove any trace evidence from the surface of the suspected tool.

2. Make several tool marks on a softer metal surface (i.e., copper or aluminum sheets) with the suspected tool (see Figure 19.6). A softer metal is used so the tool is not altered in any way.

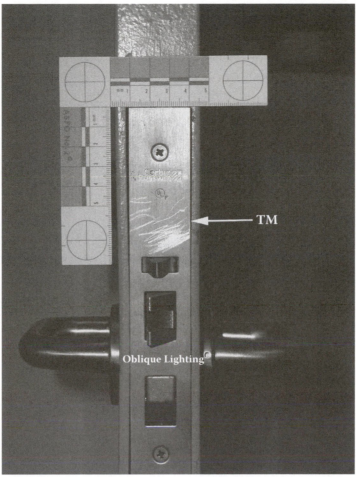

Figure 19.2
Photograph of possible tool marks on a mortise lock.

Figure 19.3
Mix the casting silicone as needed with a spatula.

Figure 19.4
Application of casting silicone onto suspected tool marks.

Figure 19.5
Gently remove the silicone cast for further examination.

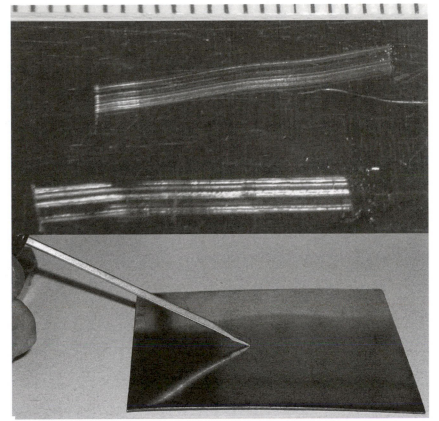

Figure 19.6
Known tool marks being made with suspected tool.

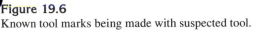

3. Attempt to simulate the manner in which the questioned tool marks were made (i.e., same angle, same degree of pressure, similar stroke, etc.).

4. Cast the known tool marks with silicone casting material.

Examination and Comparison of Questioned and Known Tool Marks

1. Typical features to be observed during a tool mark examination and comparison: class characteristics produced during the tool manufacturing process (i.e., design of the tool, shape, size, model, and accidental or individualizing marks produced randomly during everyday use).

2. Class characteristics are intentional or unavoidable characteristics that repeat during the manufacturing process and are shared by all the tools being produced. These are size, shape, pattern design, manufacturing characteristics, and wear of machinery.

3. Accidental or identifying characteristics are cuts, tears, gouges, cracks, etc. that occur randomly when a tool is used. These are often used to individualize a known tool to questioned tool marks.

4. Compare the questioned and known silicone casts under the stereomicroscope. Make all necessary measurements. Compare the known marks on the metal surface with the marks on the questioned surface. In addition, if a comparison microscope is available, compare the questioned and known marks simultaneously.

Evaluation Process

1. The questioned impression is always compared to the known (not the other way around).

2. Compare the class and accidental features of the questioned to the known.

3. Since identifying characteristics are constantly being lost and formed with use of the tool, it is possible, and likely, that characteristics will not be found on both the questioned and known impressions. Depending upon the depth and location of the characteristic, it can be eliminated in 1 day to 2 weeks of continuous use. In addition, characteristics found may have been altered in shape and size.

4. Tool mark evidence
   - Can lead to the identity of a suspect.
   - Can eliminate suspects.
   - Can determine a general or specific type of tool.
   - Can positively identify a tool (unique characteristics).
   - Can prove presence at the scene of a crime.
   - Can help reconstruct a crime scene.
   - May show involvement in the crime.
   - Can verify or disprove alibis.

Conclusions

1. For a positive identification, all class characteristics present in the questioned impression must be present in the known. In addition, one or more identifying characteristics must be present in each.

2. Positive elimination occurs when a known tool does not possess the same class features as the questioned tool marks.

3. Not suitable for comparison means not enough of the questioned mark is available.

Name _____   Date _____

Report

1. Report your findings in your notebook.

2. Justify your conclusions.

3. Make a list of matching features between the questioned and known tool marks.

Experiment  20

Glass Fractures and Direction of Force

Teaching Goals

The objectives of this lesson are to provide the student with the concepts, rationale, and procedures necessary to determine in which direction force was applied when a piece of window glass is pierced by a projectile. In addition, the student will gain hands-on experience with the handling of fragile evidence.

Background Knowledge

Glass can often be a crucial form of physical evidence used in solving crimes. Broken bottles, fractured architectural windows, smashed automobile windshields, and other types of glass objects are frequently used as physical evidence in solving murders, robberies, vehicular assaults, assaults with firearms, and many other kinds of incidents. The examination of broken glass edges can play an important role in the reconstruction of events associated with specific types of crimes.

When a glass window is struck by a projectile a fracture pattern similar in appearance to a spider's web is formed. The web pattern is composed of two primary types of fracture lines: radial and concentric.

Radial fractures branch out from the point at which the object struck the pane toward the edge of the pane. Concentric fracture lines are concentric circular fractures surrounding the point of impact. The classification of which lines are radial fracture lines and which are concentric fracture lines is fundamental to the reconstruction of which side of the window received the force that produced the resulting fractures. Therefore, close scrutiny of both of these types of fracture lines is essential before any determination of the direction of force or the sequence of impact can be determined.

Equipment and Supplies

1. Questioned broken pieces of window glass and Plexiglas as provided by the instructor
2. Thick leather gloves for handling the broken pieces of glass and Plexiglas [**CAUTION: BROKEN GLASS AND PLEXIGLAS ARE EXTREMELY SHARP! ANYONE HANDLING THESE ITEMS OF EVIDENCE MUST WEAR PROTECTIVE LEATHER GLOVES!!!**]
3. Magnifying glass and stereomicroscope

Hands-on Exercise

1. Listen carefully to the instructions for handling sharp objects safely.
2. Place the leather safety gloves on **both** of your hands before handling the glass.
3. Note the side of the window which is most soiled.
4. Examine the item you receive for radial and concentric fractures.
5. If a projectile strikes the surface of a piece of glass or plastic at a right or 90-degree angle to that surface, a cone-shaped hole is produced. Carefully examine any hole(s) in the glass pane. The peak of the cone-shaped hole is usually the point of impact.
6. Examine with a magnifying glass or a stereomicroscope the broken edge of a radial fracture for feather-shaped stress marks known as conchoidal fractures. The general rule as stated is: right, rear, radial. This rule is interpreted as follows: For a radial fracture, the force was applied to the opposite or rear side of the glass in which the conchoidal fracture forms a right angle to the pane's surface, as shown in Figure 20.1. Keep in mind that this rule is foolproof only when the radial fracture is near the point of impact, that is, closer than the first intersecting concentric fracture.

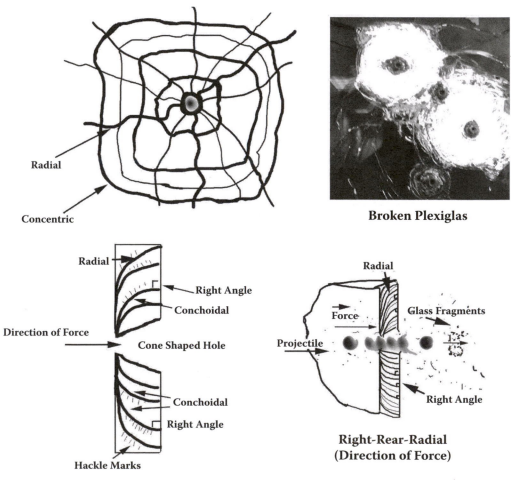

Broken Plexiglas

**Right-Rear-Radial
(Direction of Force)**

Figure 20.1
Concentric and radial fracture lines in glass.

Name _____ Date _____

Report

1. Make a sketch in your laboratory notebook of the glass specimen you received and report your data and findings.

2. Justify your conclusions.

3. Why do you think you were told to note the most soiled side of the windowpane?

4. Observe and study the diagrams in Figure 20.2.

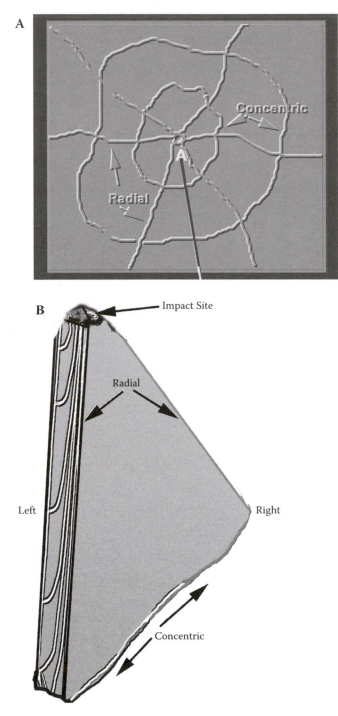

Figure 20.2
(A) A glass pane that has been penetrated by a projectile. One small section of the fracture is removed and is shown in (B).

5. Determine the direction of force from the data you collected from your observations of Figure 20.2.

6. Observe and study Figure 20.3. Is this fracture series possible? If not, why not? What is wrong? Why? Explain?

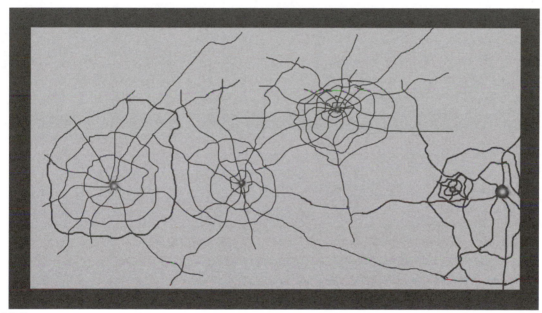

Figure 20.3
A sequence of five bullet holes fired into a pane of glass.

7. Observe and study the fracture series in Figure 20.4. Explain which fracture occurred first. Can you determine which hole was made last? Why?

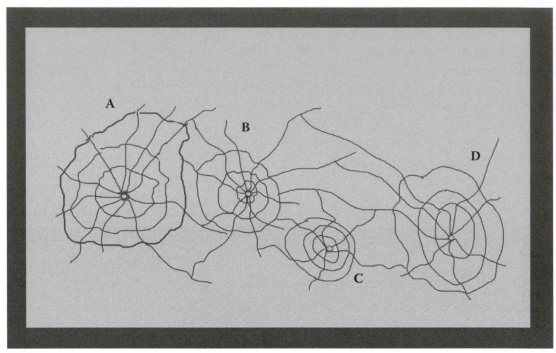

Figure 20.4
A sequence of four bullet holes fired into a pane of glass.

Experiment  22

Bloodstain Geometry (Part A)

Teaching Goals

The objective of this lesson is to allow the student to become familiar with the basic geometric shapes and patterns formed by droplets of blood when they impact various target surfaces at a 90° angle. The student will also learn the effects that various surface textures have on the geometry of blood droplets and the basic concepts and rationale employed to recognize, document, and interpret bloodstain evidence.

Background Knowledge

Blood spatter patterns are often present at the scenes of violent crimes. It is crucial that blood patterns are carefully documented and preserved in their original state so they can be examined by a blood spatter expert. Careful scrutiny of these patterns often gives valuable insight into the events of the crime. Forensic investigators must be careful to document the overall configuration of a bloodstain pattern. Details such as the location, distribution, morphology, and appearance of bloodstains and spatter patterns may be helpful in interpreting and reconstructing the crime.

Some important factors that must be recorded which are crucial to the correct interpretation of any bloodstain pattern are

- Shape of the overall pattern
- Shape of the individual blood droplets
- Absorbance properties of the substrate material
- Texture of the substrate material
- Surface characteristics of the substrate material

Some factors that can be determined from bloodstain patterns are

- Direction of travel
- Point of origin
- Impact angle
- Trajectory
- Type of weapon used
- Whether the perpetrator is right- or left-handed

Equipment and Supplies

1. Artificial blood (see instructions for preparation below)
2. A 30- to 60-mL dropper bottle or a 50-mL beaker for storing artificial blood

3. Disposable pipettes

4. Plain white typing paper

5. 12-in. square pieces of cardboard

6. 12-in. square pieces of 1/8-in. thick, smooth, clear, or white sheets of acrylic

7. 12-in. square pieces of plywood

8. Paper towels

9. Tape measure

10. Ruler

11. Meter stick

12. Protractor

Hands-on Exercise

1. Drop one drop of artificial blood straight downward onto each surface texture being tested from the height of 1, 2, 36, and 72 in.

2. The test surfaces should be placed flat on the floor and the pipette should be held perpendicular to the surface when the blood drops are released.

3. Draw a well-mixed quantity of artificial blood into the pipette and wipe off any excess blood from the outside of the pipette with a paper towel. Be careful not to include any air bubbles with the blood.

4. Make your observations while the stains are wet and after the stains have thoroughly dried.

Artificial Blood Preparation

Mixture 1: Use 4 oz of evaporated milk, 2–3 tablespoons of tomato paste, and red food dye. Mix to the viscosity of blood (use water to thin as necessary). This artificial blood mixture should be freshly made; it is only good for a few days. Store in a refrigerator.

Mixture 2: Carnation- dry milk, red food dye, and water. Mix to the viscosity of blood.

Mixture 3: White corn syrup and red food dye. Mix to the appearance of blood.

Name _____ Date _____

Report

1. Report all observations in your notebook.

2. Sketch or photograph the appearance of each specimen (see Figure 22.1).

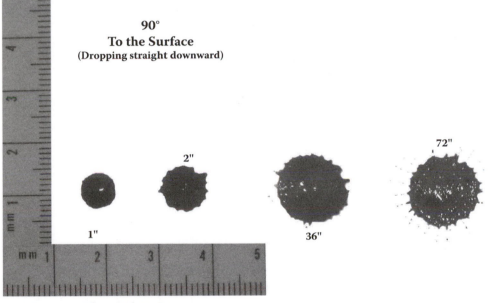

1 Drop on Plain White Paper

Figure 22.1
Patterns that are formed when one drop of fake blood is dropped straight down onto a piece of white typing paper from various heights.

3. For each surface texture being examined make the following observations:
   • Describe the texture of the surface: hard or soft, smooth or rough, porous or nonporous, absorbent or nonabsorbent.

   • Describe the edge characteristics of the resulting stains.

   • Measure the diameter of each drop (in millimeters).

   • Describe the extent of peripheral satellite spattering.

   • Discuss what effect changing the dropping heights had on the resulting stains.

   • Experiment with other surface textures such as floor tiles, carpet, and textiles.

Bloodstain Pattern Glossary

Angle of impact — The angle at which a blood drop strikes a target surface.

Back spatter — Blood that travels back toward the direction of the initiating force.

Blood spatter — The resulting pattern that is formed when a quantity of blood impacts a surface.

Cast-off blood — Blood that is collected by and then cast from a moving object during an incident.

Contact stain — The result of a bloody article coming into contact with a surface.

Point of origin — The location from which the blood originates.

Smear — The pattern left when a bloody object is wiped across a surface.

Target — The surface on which blood is deposited.

Transfer pattern — A recognizable pattern that is transferred to a surface when bloody objects come into contact with it.

Bloodstain Geometry (Part B)

Teaching Goals

The goals of this lesson are to allow the student to become familiar with the basic geometric shapes and patterns formed by droplets of blood when they impact different target surfaces at various angles. Each student will also gain additional hands-on experience while experimenting with blood smears and contact pattern transfers. Finally, each student will learn the concepts and rationale used in the study and interpretation of complex bloodstain patterns.

Equipment and Supplies

1. Artificial blood prepared in accordance with instructions in Experiment 22
2. A 30- to 60-mL dropper bottle or a 50-mL beaker and disposable pipettes
3. Plain white typing paper
4. 12-in. square pieces of cardboard
5. 12-in. square pieces of 1/8-in. thick smooth, clear, or white acrylic sheets
6. 12-in. square pieces of plywood
7. Paper towels
8. Tape measure, ruler, meter stick, and protractor

Hands-on Exercise

1. Draw artificial blood into the pipette or dropper and wipe off any excess from the outside with a paper towel. Be careful not to include any air bubbles.
2. The test surfaces should be placed at various angles to the floor. While holding the loaded pipette perpendicular to the floor, release the blood drop (see Figure 23.1).
3. Release a single drop of artificial blood straight down onto each surface texture being tested from a height of 12 in. for the following impact angles: 45°, 60°, and 70° (see Figure 23.2). Prepare one test surface for a drop of artificial blood impacting a surface at 60° from a height of 24 in. (see Figure 23.3).
4. Make your observations while the stains are wet and after the stains have thoroughly dried.
5. Repeat the experiment on several different surfaces.
6. Prepare one smear and one contact pattern on separate pieces of plain white paper (see Figure 23.4).

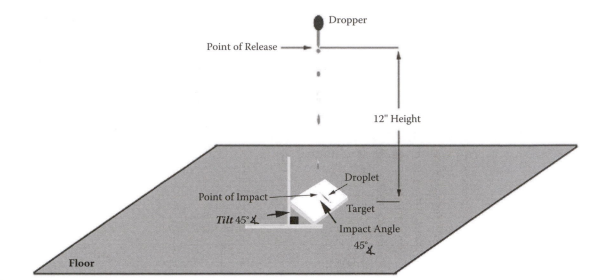

Figure 23.1
The application of the test droplets of fake blood.

12" Height on White Paper

Figure 23.2
The appearance of blood droplets after impacting a plain white paper surface at various angles from a height of 12 in.

60° to Surface

1 mm

24"

12"

**1 Drop
Plain White Paper**

Figure 23.3
The appearance of two blood droplets after impacting a piece of plain white paper at 60° from heights of 24 in. and 12 in.

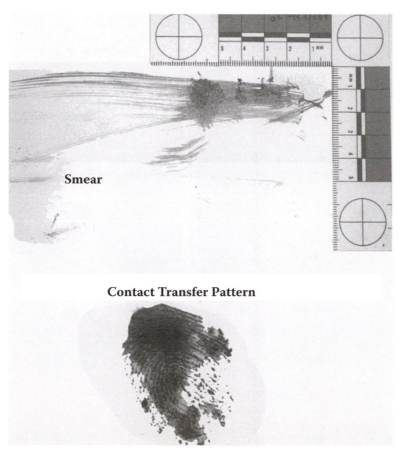

Smear

Contact Transfer Pattern

Figure 23.4
ne smear and one contact transfer blood pattern.

Name _____ Date _____

Report

1. Report all observations in your notebook.
2. Sketch or photograph the appearance of each specimen (see Figure 23.2 and Figure 23.3).

3. For each surface texture being examined make the following observations:
   - Describe the texture of the surface: hard or soft, smooth or rough, porous or nonporous, absorbent or nonabsorbent.

   - Describe the edge characteristics of the resulting stains.

- Measure the diameter and length of each pattern (in millimeters).

- Describe the extent of peripheral satellite spattering.

- Discuss what effect changing the dropping heights had on the resulting stains.

- Experiment with other surface textures such as floor tiles, carpet, and textiles.

- How do the 12- and 24-in. 60° impact angle test patterns differ?

- How do the smear and contact transfer patterns differ?

Experiment

24

Forgery Detection

Teaching Goals

The goal of this lesson is to introduce the student to basic concepts and rationale necessary to recognize, document, compare, and identify a handwritten forgery.

Background Knowledge

A large portion of the casework performed by forensic document examiners involves the identification of handwritten documents such as historical records, letters, wills, and checks. The primary process used to identify authorship involves the comparison of known handwriting with the handwriting on questioned documents.

The four basic methods used by forgers to carry out their work are tracing, freehand copying, mechanical lifting, and optical or printing reproduction. This exercise will concentrate on the detection of handwritten forgeries.

Equipment and Supplies

1. Known handwriting (script) exemplars
2. Known handwriting (print) exemplars
3. Hand magnifying glass
4. Stereomicroscope
5. Ruler
6. Protractor
7. Blue ink pens (same type used by all students)
8. Prepared known exemplar specimens
9. Prepared questioned exemplar specimens
10. Black-and-white photographic equipment
11. 3 × 4 in. white index cards

Hands-on Exercise

1. Preparation of documents. Your instructor will direct you to prepare certain handwriting exemplars and documents. Follow all instructions precisely. Do not add any identifying information to the exemplars.

   A. **Preparation of script handwriting exemplar (normal):** In your normal handwriting, write the following five items on separate index cards:

171

1. Four score and seven years ago, our fathers brought forth on this continent a new nation, conceived i
   liberty and dedicated to the proposition that all men are created equal.
2. The little sly, quick, clever, young, red fox jumped over the big, drowsy, lazy, guard dog.
3. Goodnight, goodnight! Parting is such sweet sorrow; shall I say goodnight, 'til it be morrow?
4. 111, 2222, 3333, 4444, 5555, 6666, 7777, 8888, 9999, 0000.
5. --,,//!!??() "" <> # # $$ %% && ++ [].. ;; ::.

B. **Preparation of script handwriting exemplar (disguised):** Disguise your normal handwriting and write the five items below on separate index cards:

1. Four score and seven years ago, our fathers brought forth on this continent a new nation, conceived in liberty and dedicated to the proposition that all men are created equal.
2. The little sly, quick, clever, young, red fox jumped over the big, drowsy, lazy, guard dog.
3. Goodnight, goodnight! Parting is such sweet sorrow; shall I say goodnight, 'til it be morrow?
4. 111, 2222, 3333, 4444, 5555, 6666, 7777, 8888, 9999, 0000.
5. --,,//!!??() "" <> # # $$ %% && ++ [].. ;; ::.

C. **Preparation of printed handwriting exemplar (normal):** Use your normal handwriting and print the following five items on index cards:

1. Four score and seven years ago, our fathers brought forth on this continent a new nation, conceived in liberty and dedicated to the proposition that all men are created equal.
2. The little sly, quick, clever, young, red fox jumped over the big, drowsy, lazy, guard dog.
3. Goodnight, goodnight! Parting is such sweet sorrow; shall I say goodnight, 'til it be morrow?
4. 111, 2222, 3333, 4444, 5555, 6666, 7777, 8888, 9999, 0000.
5. --,,//!!??() "" <> # # $$ %% && ++ [].. ;; ::.

D. **Preparation of printed handwriting exemplar (disguised):** Disguise your normal handwriting and print the following five items on index cards:

1. Four score and seven years ago, our fathers brought forth on this continent a new nation, conceived in liberty and dedicated to the proposition that all men are created equal.
2. The little sly, quick, clever, young, red fox jumped over the big, drowsy, lazy, guard dog.
3. Goodnight, goodnight! Parting is such sweet sorrow; shall I say goodnight, 'til it be morrow?
4. 111, 2222, 3333, 4444, 5555, 6666, 7777, 8888, 9999, 0000.
5. --,,//!!??() "" <> # # $$ %% && ++ [].. ;; ::.

E. **Preparation of questioned exemplars:** In the empty spaces of Table 24.1, sign your name as Joseph John Doe. Write in your normal freehand style as if you were signing an important document. Use the empty spaces in Table 24.2 to practice copying Joseph John Doe as you wrote it in Table 24.1.

F. **Preparation of questioned check:** Using Figure 24.1, prepare a check to cash for $1000.00. Sign it as Joseph John Doe. Endorse (also as Joseph John Doe) the reverse side of the check that appears in the figure and add your account number 20222 2842 to your endorsement.

TABLE 24.1
Model Signature Exemplars

ID#_____

Joseph John Doe	

TABLE 24.2
Copies of Signatures from Table 24.1

ID#_____

Joseph John Doe	

3. Write your name on one of the index cards provided by your instructor. Return it to the instructor along with the exemplars you prepared. The instructor will store the materials in a secure place until they are needed for the comparison and identification phases of the experiment.

4. On the day of the forgery laboratory session, you will receive one questioned exemplar from your instructor. You will be asked to identify the author of the questioned exemplar by comparing the questioned writing with the writing on each set of known exemplars.

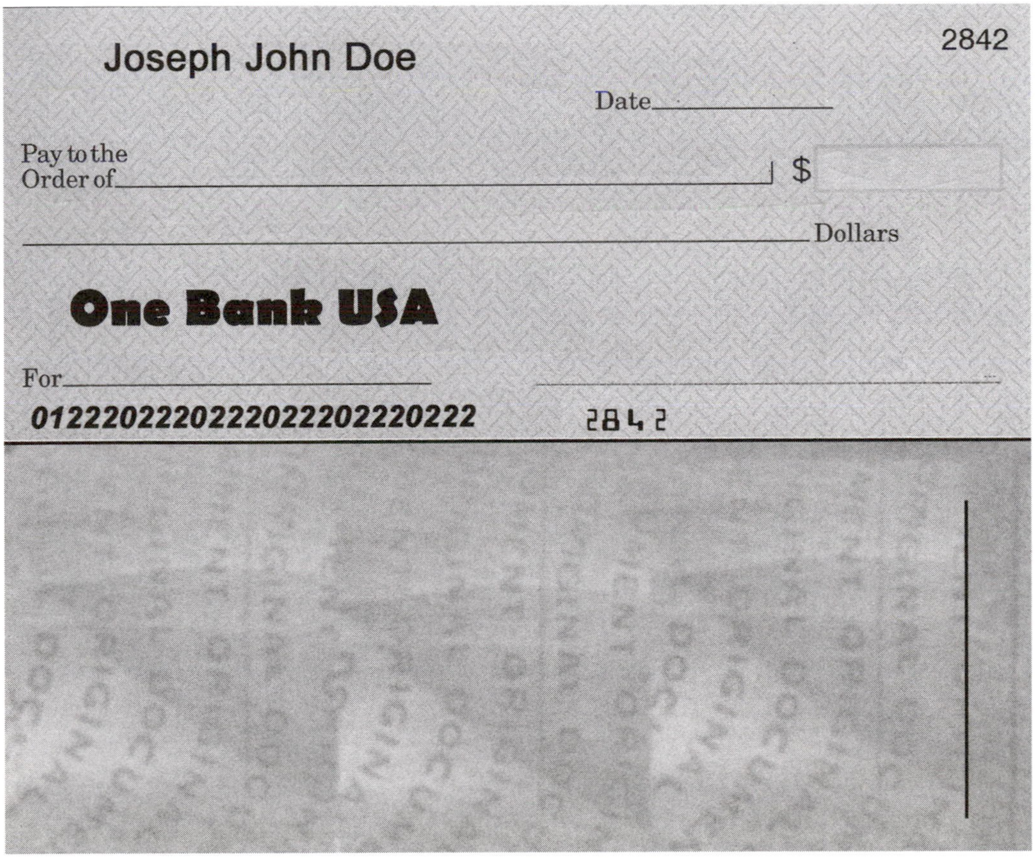

Joseph John Doe   2842

Date_____

Pay to the
Order of_____ $ _____

_____ Dollars

One Bank USA

For_____   _____

0122202220222022202220222   2842

Figure 24.1
Blank check.

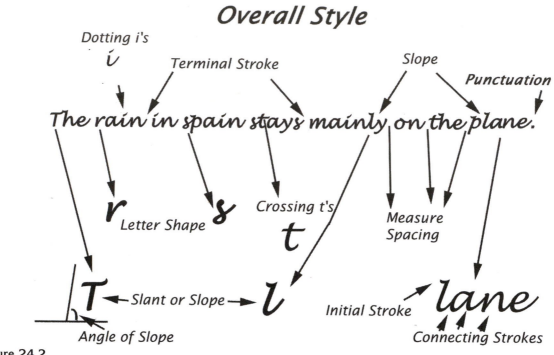

Figure 24.2
Some features to study when examining and comparing handwriting.

Name _____ Date _____

Report

1. Report your findings in your notebook. Identify the author of your questioned document with the code number on the known set of exemplars.

2. Justify your conclusions.

3. Make a list of matching features between the questioned and known handwriting.

4. Take several black-and-white photographs of the similar characteristics you used to identify the author of the questioned exemplar.

Appendix A: Important Features to Examine

- Style or overall form of writing
- Shape of letters
- Slope or slant of writing
- Spacing of letters
- Spacing of words
- Initial strokes
- Terminal strokes
- Connecting strokes

Soil Examination

Teaching Goals

This lesson will introduce the student to some of the methods and techniques used in the examination and comparison of soil specimens. Each student will learn the concepts, rationale, and procedures necessary to collect, examine, and compare forensic soil evidence.

Background Knowledge

Soil is a ubiquitous material and is often encountered in forensic casework as physical evidence. Soil from the sole of a suspect's shoe or from a vehicle can be used to connect a person or vehicle to a crime scene. Questioned soil specimens can be used to describe an environment and help reconstruct the events of a crime.

This experiment requires preliminary preparation. Two to three weeks before the experiment will be performed, each student should bring to class a 16-oz soil specimen obtained from an area around his or her home. These specimens will be used to prepare questioned and known samples.

Equipment and Supplies

1. A set of sieves: 20-, 40-, 60-, 80-, 100-, 150-, 200-mesh sizes and trap
2. Triple beam balance
3. Six test tubes (6-in.) and six 100-mL beakers
4. pH or litmus paper
5. Distilled water
6. Munsell soil color charts
7. Disposable plastic spot plates
8. Wax pencil or marker
9. Test tube rack
10. Plastic spot plates, 2 × 3 in.
11. Questioned and known soil specimens provided by the instructor

Hands-on Exercise

1. Obtain one questioned and four known soil specimens from your instructor.
2. Air dry and weigh each 100-mL beaker. Record the weight of each beaker on the Soil Data Weight Sheet below (Table 25.1). Label one beaker Q1 (Q = questioned) and the remainder K1 (K = known) through K4.

TABLE 25.1
Soil Data Weight Sheet

Specimen	Beaker Wt.	Soil Wt.	Total Wt.	Wt. SF 20	Wt. SF 40	Wt. SF 60	Wt. SF 80	Wt. SF 100	Wt. SF 150	Wt. SF 200
Q1										
K1										
K2										
K3										
K4										

3. Place each specimen into its corresponding beaker; for example, the questioned specimen should be placed in the Q1 beaker.

4. Weigh each specimen and record its weight in Table 25.1.

5. Sieve each specimen using the sieves of different meshes, starting with size 20. After sieving, place the sieved fraction on a piece of weighing paper. Weigh the fraction and record its weight in Table 25.1.

6. Clean the sieve. Repeat the sieving and weighing steps for each specimen and sieve size.

7. Place a small quantity of each questioned and known soil sample (use the 150-mesh fraction) into individual wells of the spot plates and mark for identification. Place a similar size fraction of the known specimen into one of the top wells as depicted in Figure 25.1.

8. Figure 25.1 shows how you compare the colors of the questioned and known specimens. Compare the colors of your soil specimens while they are illuminated by sunlight. Note the results in Table 25.2.

9. More precise comparison can be achieved by using a Munsell soil color chart (see Figure 25.2). To order a chart, call 1-800-622-2384.

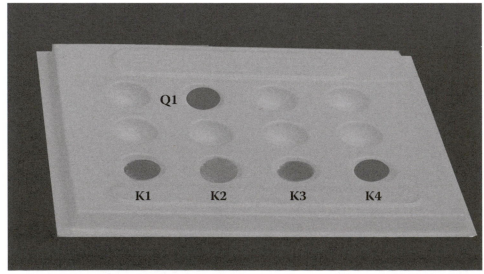

Figure 25.1
A color comparison of the questioned and known soil specimens.

TABLE 25.2
Color Comparison Table

Specimen	Soil Color (Same or Different)
Q1	
K1	
K2	
K3	
K4	

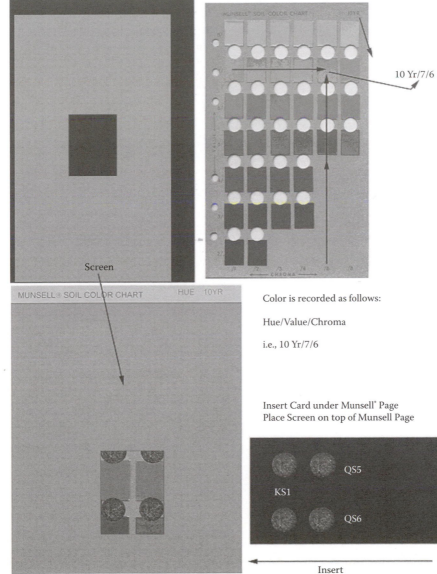

10 Yr/7/6

Color is recorded as follows:

Hue/Value/Chroma

i.e., 10 Yr/7/6

Insert Card under Munsell® Page
Place Screen on top of Munsell Page

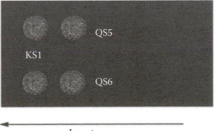

Insert

Figure 25.2
Use of Munsell soil color charts to determine a specimen's color.

TABLE 25.3
Soil pH Comparison Table

Specimen	Soil's pH or Acid-Basic
Q1	
K1	
K2	
K3	
K4	

10. Place a small quantity of each soil specimen into a labeled test tube. Add 3 to 4 mL of neutral, deionized, distilled water into each test tube and shake well. Place each tube in the rack and allow the soil specimens to settle. With a disposable plastic pipette, draw out a small quantity of water from each test tube and apply it to the pH paper. Determine the pH of each specimen and record the results in Table 25.3 and the Report section below.

Name _____ Date _____

Report

1. Prepare a chart that contains all the data you collected for each soil specimen. Based on the chart, determine whether any of the questioned samples has a common origin with the known specimen.

2. Justify your conclusions.

3. List the similar and dissimilar features of the questioned and known specimens.

4. What is soil?

5. What is the composition of soil?

6. How many minerals are normally contained in an average soil specimen?

7. Could any of the questioned samples be eliminated from the vegetation present in the 20-mesh fraction?

8. Were you able to eliminate any questioned specimens by comparing their soil fraction?

9. Were you able to eliminate any questioned specimens by color comparison?

10. What were the pH values of all your specimens? Were any questioned specimens eliminated after pH comparison?

Experiment **26**

Forensic Odontology 1
Is It a Bite Mark?

Teaching Goals

The student will learn the rationale used by forensic dentists when examining human remains. The student will become acquainted with human dentition and terminology, and will learn how to differentiate and identify human teeth.

Background Knowledge

Forensic dentistry (odontology) involves both the identification of human remains from surviving dentition, and the identification and comparison of bite mark evidence made during the commission of a crime. Techniques used in forensic dentistry for the identification of human remains are often helpful in situations where the traditional methods of identification (such as fingerprints, sole prints, photographic recognition, blood or tissue groups) are not possible because of the condition of the remains. In these circumstances, forensic dentists often use the past dental records of the suspected individual to prove or disprove his or her identity. Old x-rays and charts made during periodic visits to the dentist are often used to this end.

Bite marks left at the scene of a crime on a victim or other item are often used to prove or disprove the involvement of a suspect in a particular crime. Normally photographs, casts, and tracings of the suspected bite marks are directly compared to a cast made of the suspect's teeth. Since this type of evidence involves a physical match between a questioned bite mark and a known set of teeth, it can be used to prove conclusively that a person did or did not make a questioned mark, thereby helping to establish a person's guilt or innocence.

The following exercise is designed to help introduce the student to bite mark examinations and comparisons.

Equipment and Supplies

1. Diagram showing anatomy and terminology used for the jaws and teeth (Figure 26.1a)
2. Diagram showing anatomic terms and the position of each tooth in the jaw (Figure 26.1b)
3. A diagram of practice bite marks (Figure 26.1f)
4. Questioned human bite marks and tool marks (Figures 26.2a–f) and the marks on various items supplied by your instructor
5. Known bite mark patterns (Figures 26.3a–c)
6. Sketching equipment, pencils, tracing paper, and graph paper

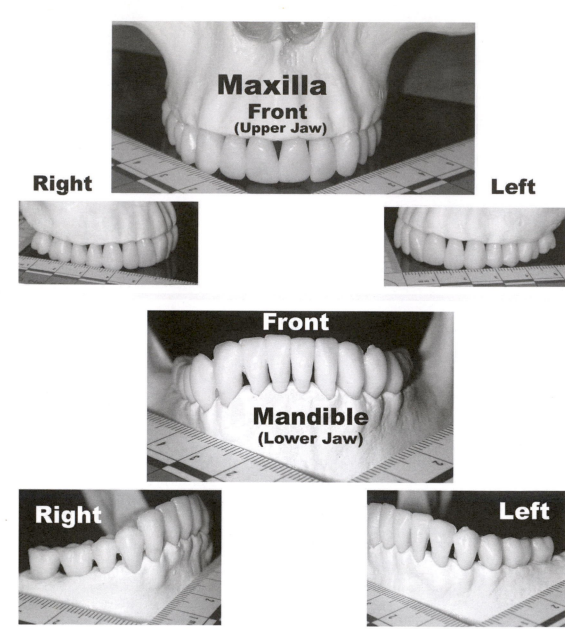

Figure 26.1a
Anatomic terms for the upper (maxilla) and lower (mandible) jaw.

7. Magnifying glass and stereomicroscope

8. Ruler, micrometer, and protractor

Hands-on Exercise

1. **Study of Human Dentition and Nomenclature**

   a. Study the information given in Figure 26.1a.

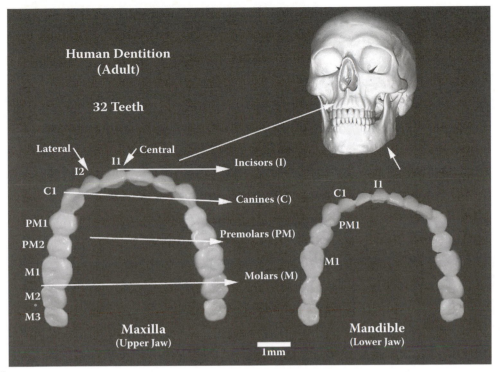

Figure 26.1b
Terms and positions for each tooth in the human jaw. The third molar (M3), also known as the wisdom tooth, is often absent.

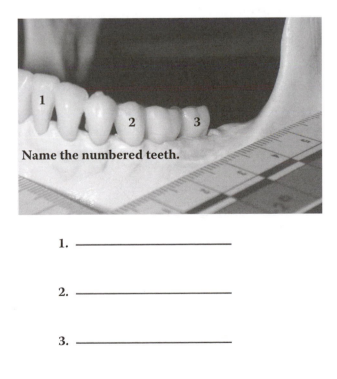

1. _____

2. _____

3. _____

Figure 26.1c

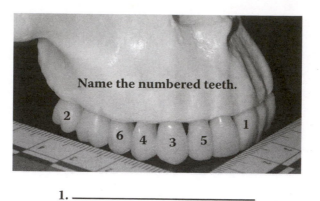

1. _____

2. _____

3. _____

4. _____

5. _____

6. _____

Figure 26.1d

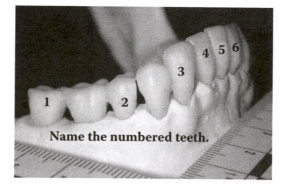

1. _____

2. _____

3. _____

4. _____

5. _____

6. _____

Figure 26.1e

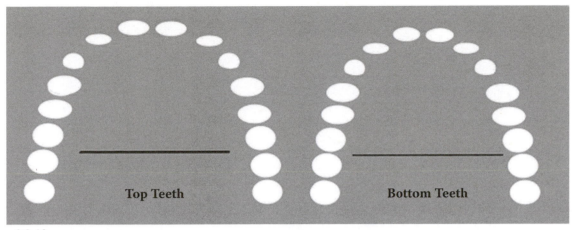

Figure 26.1f

b. Continue studying the anatomic terms used for each tooth as well as the position of each tooth in the upper and lower jaw (see Figure 26.1b).

c. Practice naming the individual teeth shown in Figures 26.1c–e.

d. Fill in the correct abbreviation for each tooth position shown in Figure 26.1f.

2. **Study of Known Impressions**

a. Examine the questioned marks in Figures 26.2a–e and the marks on the items supplied by your instructor. The marked items received from your instructor should be examined *in situ* using oblique light.

b. Figures 26.3a–c are human bite mark impressions made in wax. Figure 26.3b and Figure 26.3c illustrate the types of linear and angular measurements that can be made when examining questioned marks.

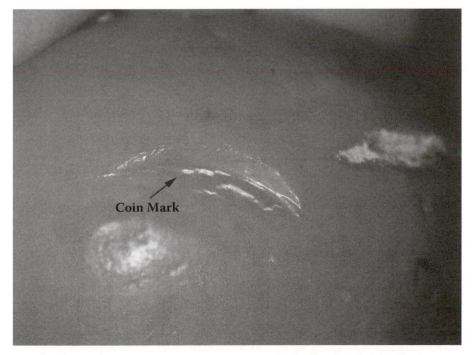

Figure 26.2a

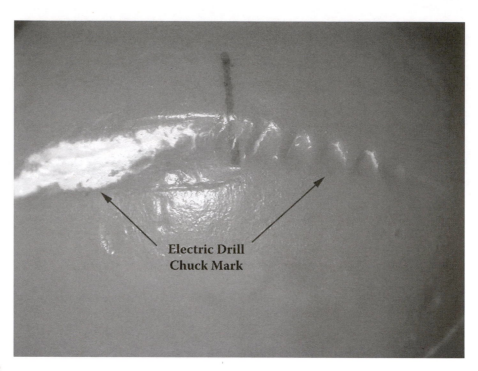

Figure 26.2b

c. Sketch or trace the marks in Figures 26.2a–e and Figure 26.3a in your laboratory notebook. Carefully practice making linear measurements of the questioned marks with a ruler and micrometer. Make angular measurements with a protractor on these same figures. Record all of your observations in Table 26.1. (Tracing is accomplished by placing a piece of tracing paper directly over the questioned mark and carefully tracing the mark's outline with a pencil.)

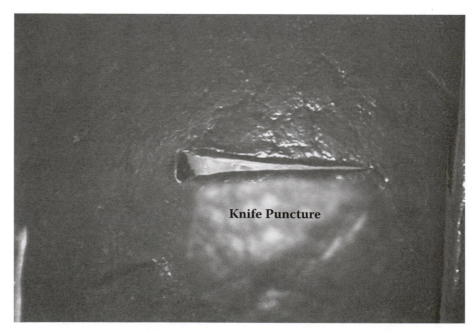

Figure 26.2c

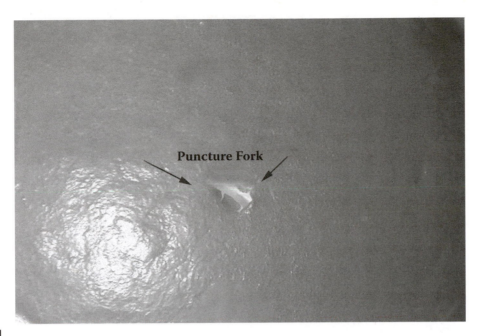

Figure 26.2d

3. **Examination of Questioned Bite Marks**

   a. Typical features to be observed during a bite mark examination are as follows: overall appearance of the bite mark, spacing distances between teeth, angles of misaligned teeth, shape of each tooth, number of teeth present, and number of missing teeth.

   b. Determine if the marks in Figures 26.4a–c are bite marks or tool marks.

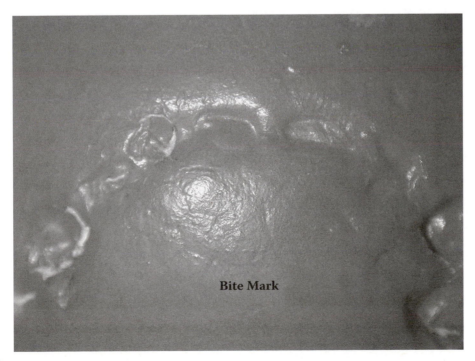

Figure 26.2e

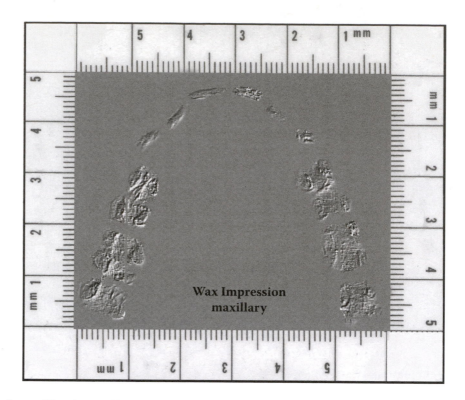

Figure 26.3a
Wax impression of a maxillary human bite mark.

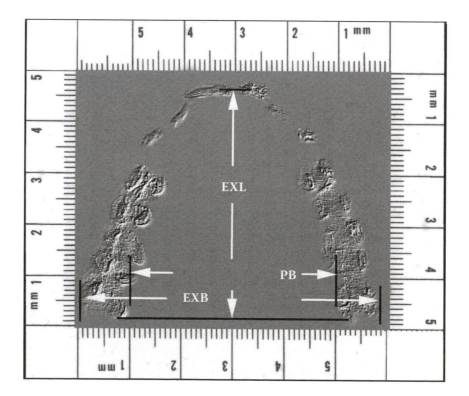

Figure 26.3b
Linear measurements of bite marks: external breadth (EXB), palate breadth (PB), and external length (EXL) are but a few of the linear measurements one can make when examining bite marks.

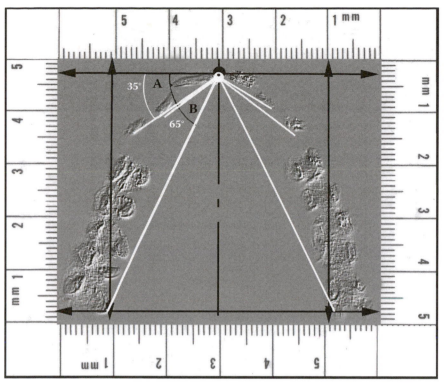

Figure 26.3c
ngular measurements of bite marks.

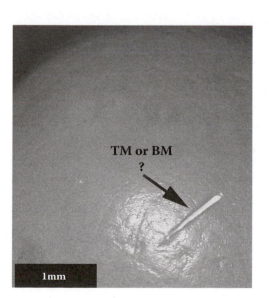

Figure 26.4a
termine if this is a tool or bite mark. Why?

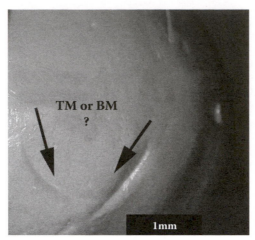

Figure 26.4b
Determine if this is a tool or bite mark. Why?

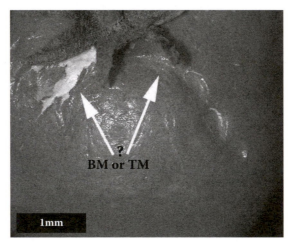

Figure 26.4c
Determine if this is a tool or bite mark. Why?

TABLE 26.1
Bite Mark Measurement Data

Specimen	Linear measurements	Angular measurements	Type of mark	Comments
Figure 26.4a				
Figure 26.4b				
Figure 26.4c				

Name _____ Date _____

Report

1. Record all observations in your notebook.

2. Draw a rough sketch in your notebook.

3. Report your findings.

4. Justify your conclusions.

Glossary of Terms for Human Bite Marks

Bite mark — Any mark or series of marks made by teeth when they are applied to a softer surface with pressure.

Canine — Teeth with pointed cusps designed for tearing and cutting.

Dentition — The kind or type, number, and arrangement of teeth of humans and animals.

Incisor — A tooth designed for cutting.

Maxillary — The upper stationary jaw of the human skull.

Mandible — The lower movable jaw of the human skull.

Molars — Teeth with very broad occluded surfaces and multiple cusps designed for grinding and macerating food.

Odontologist — Dentist who identifies unknown human remains from their surviving dentition and who help solve crimes by examining bite mark evidence.

Premolars — Teeth with broad surfaces and multiple cusps designed for grinding and macerating food.

Experiment **27**

Forensic Odontology 2
Who Made the Questioned Bite Mark?

Teaching Goals

The student will learn the rationale used by forensic dentists (odontologists) when comparing questioned bite marks with known dental casts. In addition, each student will gain hands-on experience doing examinations and comparisons of realistic bite mark evidence.

Background Knowledge

Odontologists are often asked to examine bite marks found on a victim or some other item of physical evidence. Typically photographs, casts, or tracings of the questioned bite mark are directly compared to casts made of the suspect's teeth. Since this type of evidence involves a physical match between a questioned bite mark and a known set of teeth, it can be used to prove conclusively that a person did or did not make a questioned mark, thereby helping to establish a person's guilt or innocence.

The following exercise is designed to help introduce the student to bite mark examinations and comparisons.

Equipment and Supplies

1. Photographs of questioned bite mark exemplars
2. Photographs of known dental stone casts made by a professional dentist*
3. Bite mark profiles made in wax from casts
4. Sketching equipment, pencils, tracing paper, and graph paper
5. Magnifying glass and stereomicroscope
6. Ruler, micrometer, and protractor

Hands-on Exercise

In addition to the criteria discussed in Experiment 26, factors such as bite variation (Figure 27.1), missing teeth (Figure 27.2), tooth rotation, misaligned teeth, odd-shaped teeth, and the number of cusps in each molar (Figure 27.3) are also used in any assessment of a questioned bite mark.

---

* Dental stone casts courtesy of Dr. Michael J. Mistretta, DDS, Massapequa, New York.

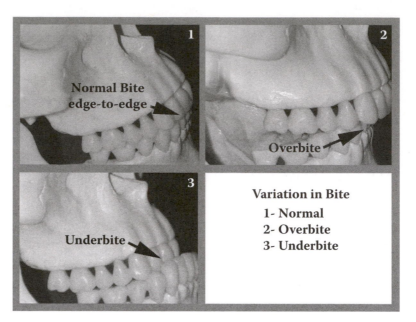

Figure 27.1
Variations in bite.

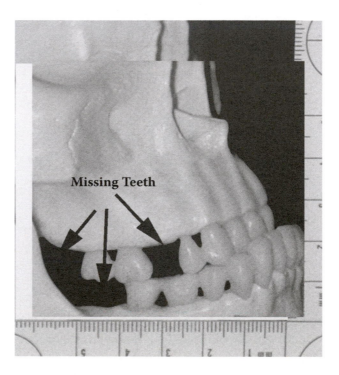

Figure 27.2
Missing teeth.

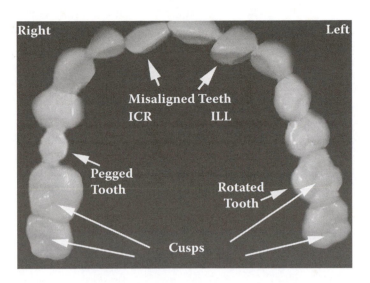

Figure 27.3
Shown are misaligned teeth, rotated teeth, pegged teeth, and several molars with various numbers of cusps.

Examination of Knowns

Examine, measure, contrast and compare the known casts and wax profiles given in Figures 27.4a,b and Figures 27.5a,b. These figures are in life size or 1:1 scale. Figure 27.6 depicts the measurements that can be made on a cast and its corresponding wax profile. Place all of the data made of the knowns in Table 27.1.

Some common features to be observed during a bite mark examination are as follows:

- Overall appearance of the mark
- Spacing distances between the teeth
- Angle of misaligned teeth
- Shape of each tooth

- Missing teeth
- Number of teeth present
- Number of missing teeth

TABLE 27.1
Data Obtained from the Knowns in Figures 27.4 to 27.7

Specimen	Linear Measurements	Angular Measurements	No. of Teeth	No. of Missing Teeth	No. of Cusps	Mark Type	Comments
K1							
K2							
K3							
K4							
K5							
K6							
K7							
K8							
K9							
K10							
K11							
K12							

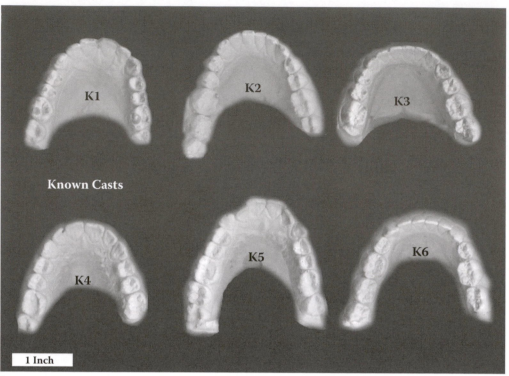

Figure 27.4a
Known dental stone casts. Scale ≈ 1:1.

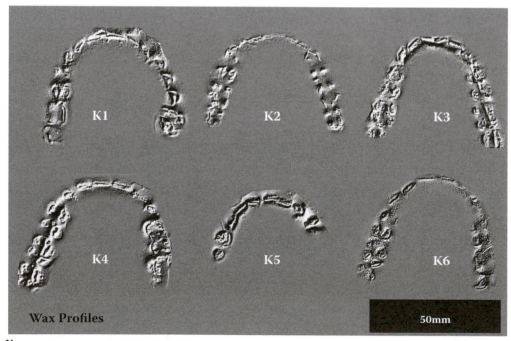

Figure 27.4b
Known wax profiles of casts. Scale ≈1:1.

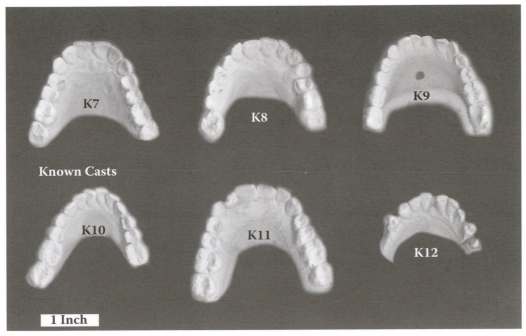

Figure 27.5a
Known dental profiles of casts. Scale ≈1:1.

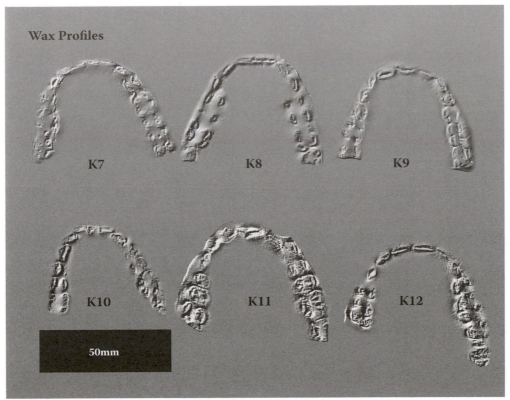

Figure 27.5b
Known wax profiles of casts. Scale ≈ 1:1.

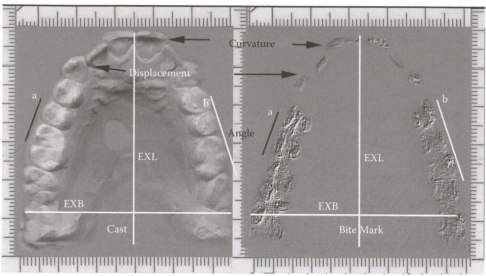

Figure 27.6
A dental stone cast and its wax profile.

Examination of Questioned Marks

Using the knowledge, skills, and abilities gained from Experiments 26 and 27, examine each of the questioned marks given in Figure 27.7 to Figure 27.13. Document all measurements made of the questioned marks in Table 27.2.

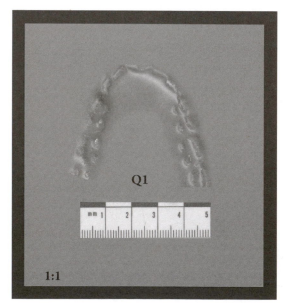

Figure 27.7
Is Q1 a bite mark? Scale ≈ 1:1.

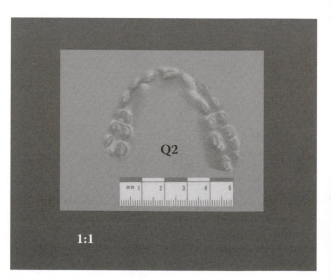

Figure 27.8
Is Q2 a bite mark? Scale ≈ 1:1.

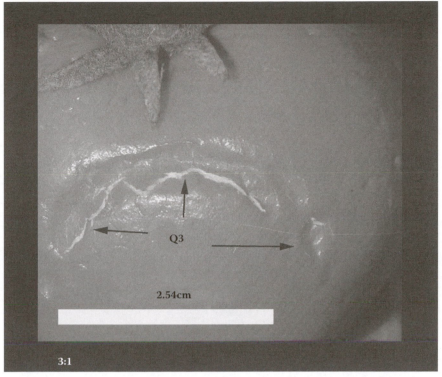

Figure 27.9
Is Q3 a bite mark? Scale ≈ 3:1.

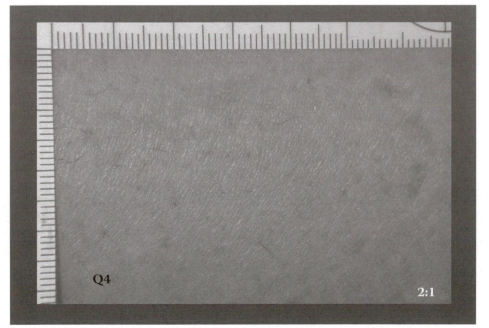

Figure 27.10
Is Q4 a bite mark? Scale ≈ 2:1.

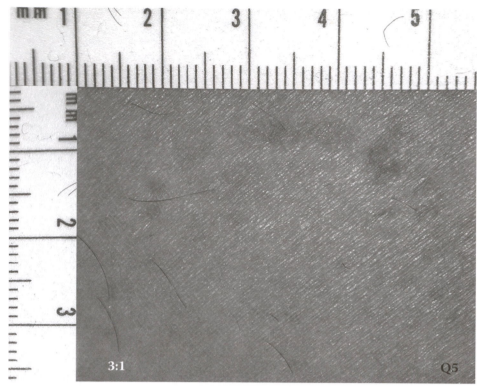

Figure 27.11
Is Q5 a bite mark? Scale ≈ 3:1.

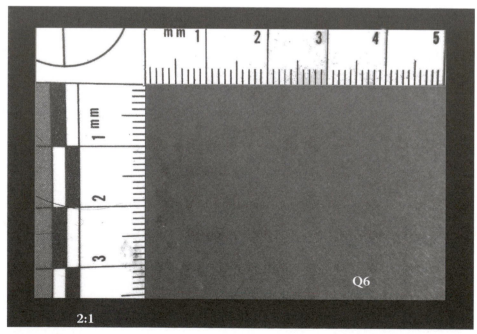

Figure 27.12
Is Q6 a bite mark? Scale ≈ 2:1.

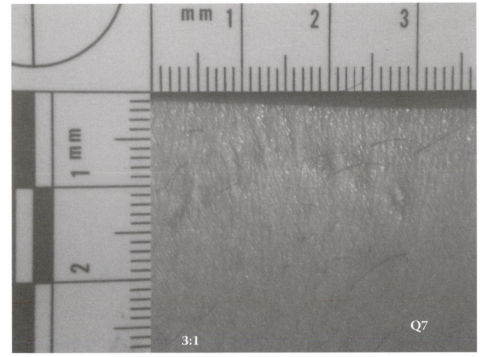

Figure 27.13
Is Q7 a bite mark? Scale ≈ 3:1.

TABLE 27.2
Data Obtained from the Questioned Marks in Figure 27.7 to Figure 27.13

Specimen	Linear Measurements	Angular Measurements	No. of Teeth	No. of Missing Teeth	No. of Cusps	Mark Type	Comments
Q1							
Q2							
Q3							
Q4							
Q5							
Q6							
Q7							

Name _____ Date _____

Report

1. Determine which of the questioned marks (if any) are bite marks. Explain your conclusions. Show all data, drawings, sketches, and tracings.

2. Compare and contrast any questioned bite marks with the known bite mark casts, photos, and wax profiles (see Figure 27.4 to Figure 27.6).

3. If possible, determine which knowns made the questioned bite marks.

4. Prepare a report, court display, and data tables to demonstrate your conclusions.

Experiment

28

Forensic Archeology
Search for Human Habitation and Remains

Teaching Goals

The student will learn the basic scientific rationale and techniques used by forensic archeologists when searching a location for evidence of human habitation, activity, and remains. Each student will learn how to locate items of archeological interest, document their position relative to the overall scene, and collect and preserve each item of interest.

Background Knowledge

The methods of archeology are often used by forensic pathologists, crime scene investigators, forensic scientists, forensic anthropologists, and forensic odontologists for the recovery of buried human and animal remains and artifacts. Archeologists use a wide range of procedures and collecting techniques for the retrieval of remains from past events, as well as for their scientific interpretation and use in the reconstruction of past events. Therefore it is important for anyone involved in the collection of physical evidence to acquire at least a rudimentary knowledge of the fundamental methods and practices used by an archeologist. The following exercise is designed to introduce the student to the basic methods of forensic archeology.

Equipment and Supplies

1. Assorted plastic human bones
2. Assorted artifacts
3. Evidence tape
4. Mason's string
5. Flagged stakes
6. Numbered cones
7. Graph paper
8. Ruler, micrometer, and protractor
9. Tape measure
10. Sieves
11. Compass
12. Scissors

Figure 28.1
A prepared site.

13. Level
14. Trowel
15. Shovels
16. Buckets
17. Cleaning brushes
18. Tooth brush
19. Metal detector
20. Packing material
21. Specimen boxes
22. Brown paper bags, various sizes
23. 35 mm camera equipment
24. Prepared archeological site (Figure 28.1)

Hands-on Exercise

1. Form a cooperative learning group of 12 to 20 students:
   a. Lead forensic archeologist (1)
   b. Lead forensic anthropologist (1)
   c. Photographers (2)
   d. Topographer (1–2)
   e. Diggers (4–6)
   f. Transporters (2–4)

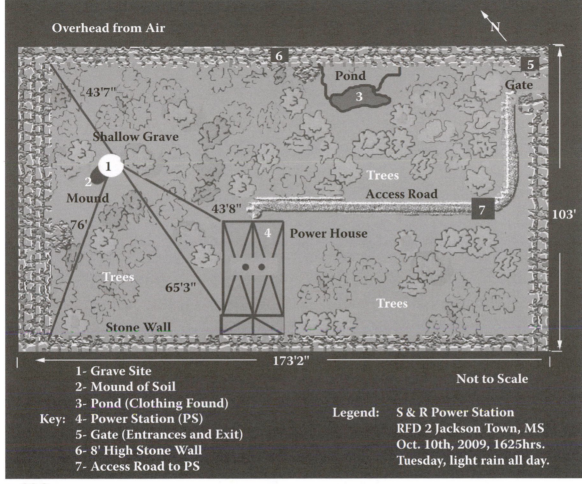

Figure 28.2
Map of overall site.

g. Groups of two for sieving soil specimens (2)

h. Record keepers (2)

i. Specimen packers (2)

2. Assign individuals to their respective task (students may volunteer for various tasks).

3. Examine the site prepared by your instructor.

4. Develop a plan for processing the site:

a. Photograph the pristine site.

b. Draw a map of the overall site (Figure 28.2).

c. Carefully scrutinize the outer perimeter of the grave.

d. Search for possible physical evidence such as shovels, footprints, tire tracks, or drag marks.

e. Document, collect, and package any physical evidence.

f. Rope off the grave site and prepare a grid with mason's string (Figure 28.3).

g. Photograph and sketch the grave site.

h. Accurately measure and document the grave site.

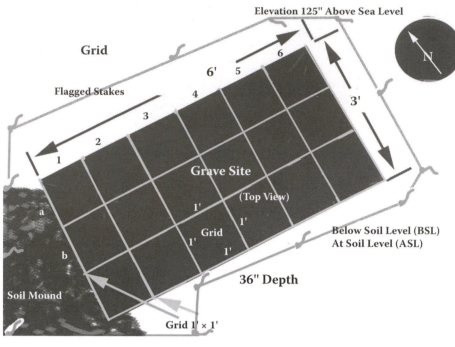

Figure 28.3
Grave site map with grid.

    i. If available, scan the area with a metal detector. Mark any potential hits with a flagged stake or cone.

    j. Carefully remove the top layer of soil while searching for items of physical evidence such as bones, teeth, tissue, hair, insects, insect larva, weapons, bullets, projectiles, or clothing.

5. Gradually work your way down to the bottom of the grave site while carefully searching for, documenting, and retrieving possible items of physical evidence (Figure 28.4 to Figure 28.6)

6. Gently remove the layers of soil 1 to 2 in. at a time while working your way to the grave's bottom.

7. Clean all artifacts with a soft brush.

8. Package each artifact in a paper bag or box. Mark and seal for identification.

9. Collect loose soil in buckets and sieve. Collect any artifacts and place them in an appropriate storage container. Mark for identification and seal each container for later examination in the laboratory.

10. Sieve all soil specimens removed from the grave for additional artifacts. Collect and document all items.

11. Package all items of physical evidence and remove to the laboratory for examination and evaluation.

Slice 1 - Surface Level

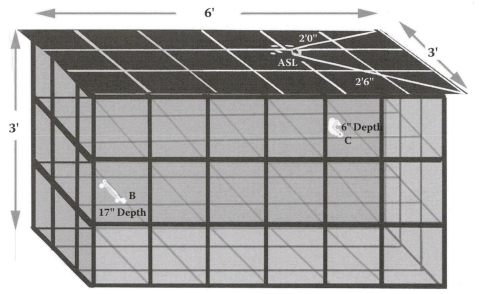

Figure 28.4
Top portion or surface level of grave site.

Slice 2 - 6" Below Surface Level

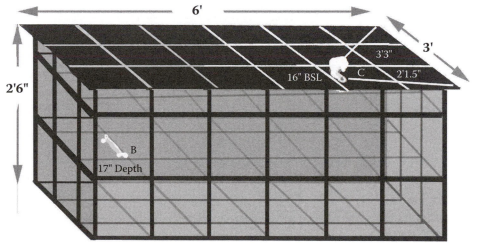

Figure 28.5
Six inches below surface level.

Slice 3 - 6" Below Surface Level

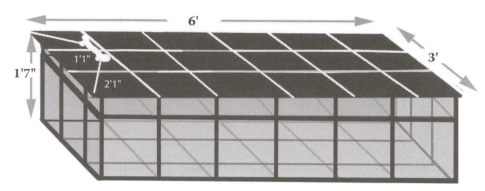

Figure 28.6
Seventeen inches below surface level.

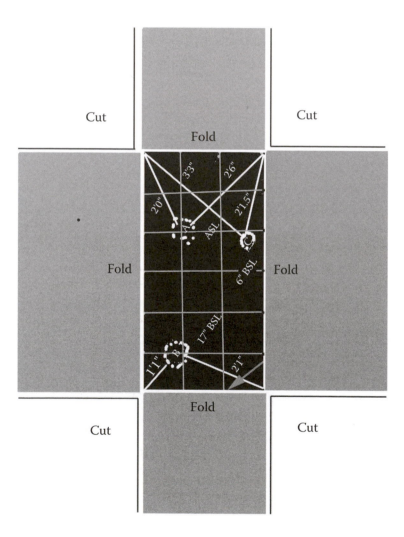

Figure 28.7
Top view of the site.

3D - Grave Site

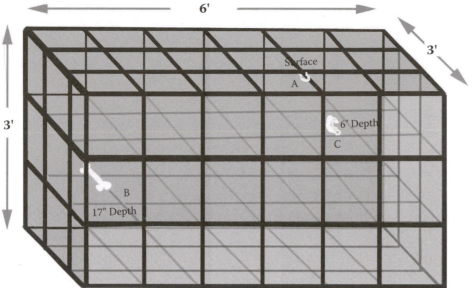

Figure 28.8
Three-dimensional view of the grave site with grid.

Name _____ Date _____

Report

1. Prepare a diagram showing the entire site with all the items of physical evidence collected at the site as well as their relationship to one another. Use Figure 28.2 to Figure 28.6 and use Figure 28.7 and Figure 28.8 as a guide.

Experiment **29**

Forensic Anthropology 1
Basic Human Osteology

Teaching Goals

The student will learn the basic scientific rationale and techniques used by forensic anthropologists when examining human bones. In addition, each student will learn how to identify human bones.

Background Knowledge

Anthropology is a science dedicated to the study of humans and their cultures. Forensic anthropology (a subdiscipline) concerns itself primarily with the medicolegal study of human remains, in particular human osteology. Forensic anthropologists are frequently called upon by police investigators to identify and interpret human remains during death investigations. The forensic anthropologist is often asked: Are these bones human? What was the person's sex? How old was the person? What was the cause of death? The following exercise is designed to introduce the student to the basic methods of forensic anthropology.

Equipment and Supplies

1. Diagram showing anatomy and anatomic terms of a human skeleton along with each bone's position (see Figure 29.1)
2. A basic plastic model of a full-size human skeleton or single full-size plastic bones
3. Sketching equipment, pencils, tracing paper, and graph paper
4. Magnifying glass and stereomicroscope
5. Ruler
6. Micrometer
7. Caliper
8. Protractor

Hands-on Exercise

1. Study the information presented in Figure 29.1. Note the names of the various skeletal elements.
2. Study the human skeletal elements presented in Figure 29.2 to Figure 29.11 and the skeletal model or bones provided by your instructor.

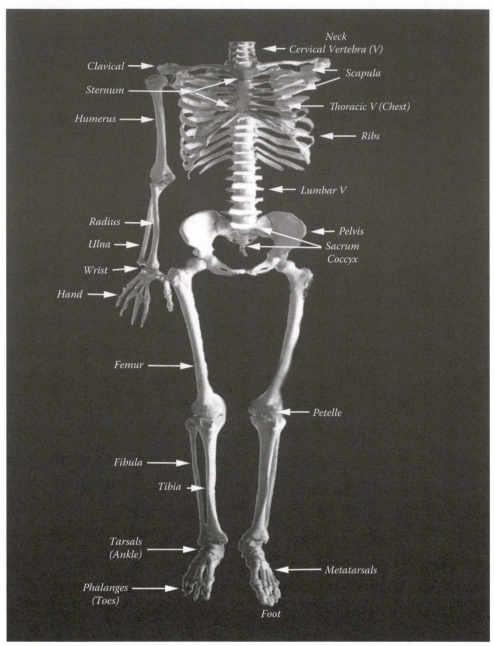

Figure 29.1
The basic human skeleton (minus skull).

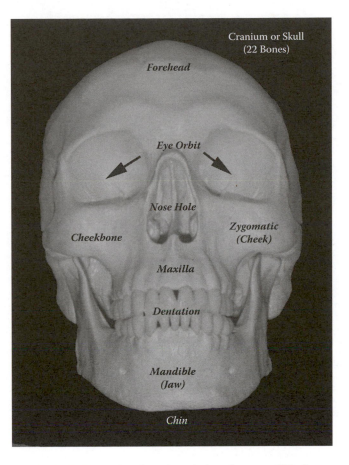

Figure 29.2a
Frontal view of an adult human male skull. The estimation of sex is based on mandible features. (Adapted from W. M. Bass, pp. 87–88.)

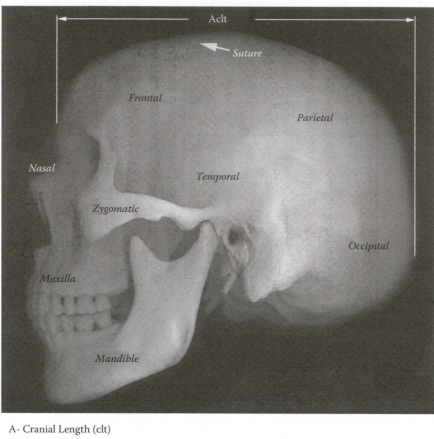

A- Cranial Length (clt)

$$\text{Cranial Index} = \frac{\text{Cranial Width (cwt)} \times 100}{\text{Cranial Length (clt)}}$$

Figure 29.2b
Some common measurements made of the skull. These measurements are used by anthropologists to help estimate the age and sex of a person at the time of death. (Measurements adapted from W. M. Bass, pp. 78–82.)

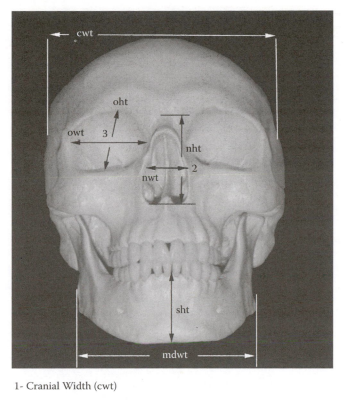

1- Cranial Width (cwt)

2- Nasal Index = $\dfrac{\text{Nasal Width (nwt)} \times 100}{\text{Nasal Height (nht)}}$

3- Orbit Index = $\dfrac{\text{Orbit Height (oht)} \times 100}{\text{Orbit Width (owt)}}$

Figure 29.2c
Lateral view of the skull. Note the measurement and calculation of the cranium. This measurement is used by anthropologists to help determine if a skull is from a human or other primate and to estimate the age and sex of the individual. (Measurements adapted from W. M. Bass, pp. 76–77.)

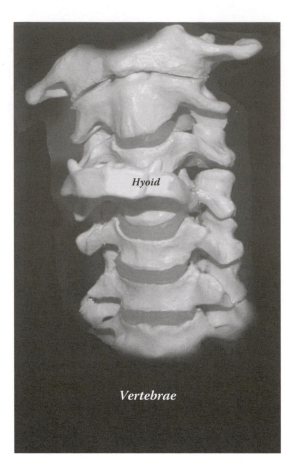

Figure 29.3
Frontal view of the first six vertebrae and the position of the hyoid bone. The hyoid bone is noted because it is often fractured during strangulation, thus it serves as a possible indicator of cause of death.

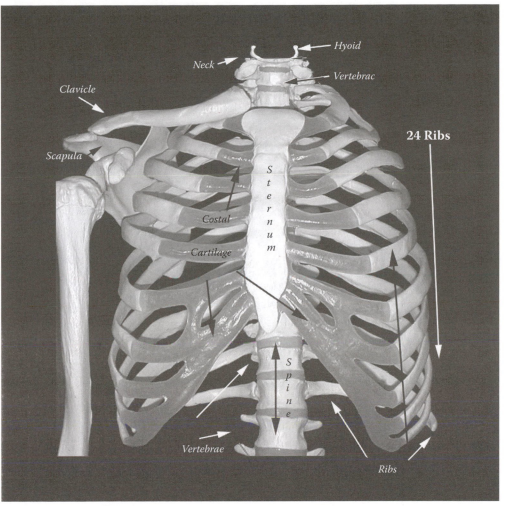

Figure 29.4
Frontal view of the thorax or rib cage and its structural elements. Note the position of the clavicle (collarbone) and the scapula or shoulder blade.

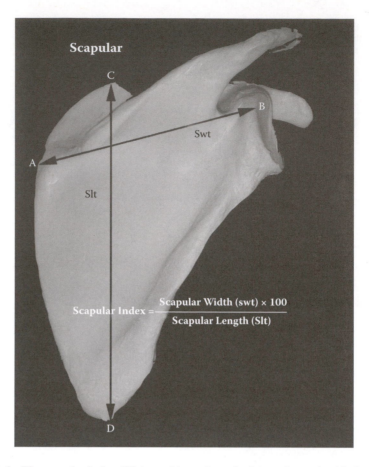

Figure 29.5
The scapula or shoulder blade. The scapular index (SI) is used by anthropologists to determine if a bone is from a human (*Homo sapiens*) or some other mammal. Measurements are made of the maximum width and length of the specimen scapula.

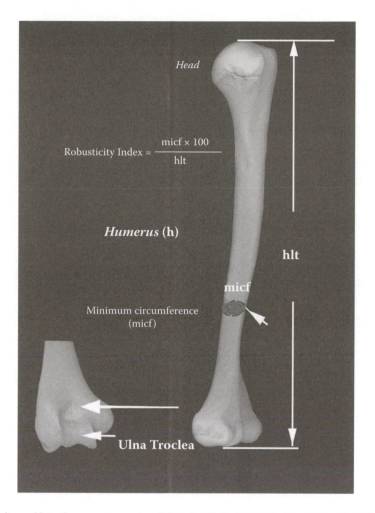

$$\text{Robusticity Index} = \frac{\text{micf} \times 100}{\text{hlt}}$$

Head

Humerus (h)

hlt

micf

Minimum circumference
(micf)

Ulna Troclea

Figure 29.6
The humerus or upper arm bone. Note the measurements of the robusticity index used to express the relative size of the shaft. The ulna trochlea is a characteristic feature of the humerus. Measurements are made of the minimum circumference and the maximum length of the humerus.

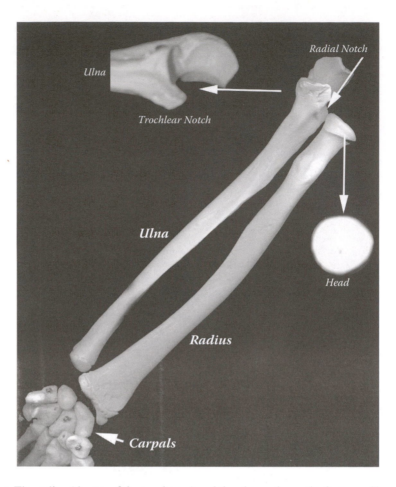

Figure 29.7
The bones of the forearm. The radius (shorter of the two bones) and the ulna make up the forearm. Shown are characteristic features of both bones.

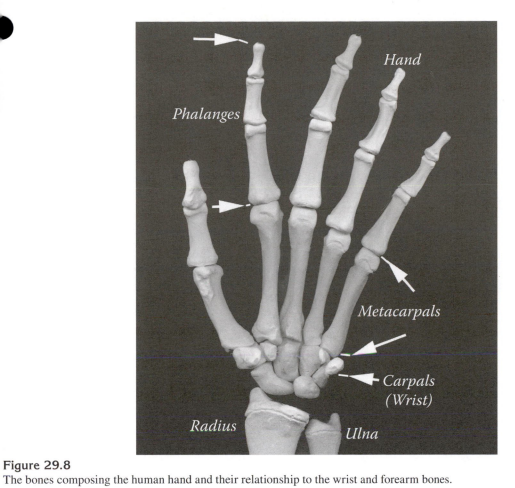

Figure 29.8
The bones composing the human hand and their relationship to the wrist and forearm bones.

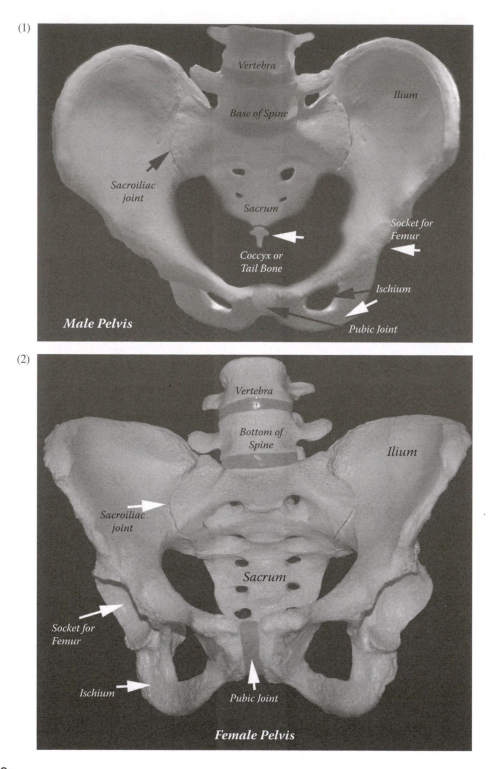

(1)

Vertebra

Ilium

Base of Spine

Sacroiliac
joint

Sacrum

Socket for
Femur

Coccyx or
Tail Bone

Ischium

Pubic Joint

Male Pelvis

(2)

Vertebra

Bottom of
Spine

Ilium

Sacroiliac
joint

Sacrum

Socket for
Femur

Ischium

Pubic Joint

Female Pelvis

Figure 29.9a
The pelvis of a male (1) and female (2) adult. The pelvis is often used to help determine the sex of an individual.

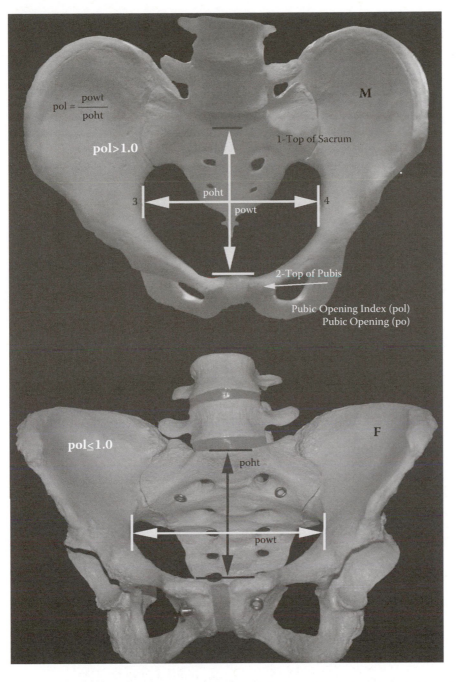

Figure 29.9b

The pelvis of a male (M) and female (F) adult. Note the size of the hole in the middle of the pelvis. At birth, the baby must pass through the hole in the mother's pelvis. Thus a female's pelvis requires a larger pelvic opening than does a male's. In general, a pelvic opening (pol) given by powt/poht greater than 1.0 indicates a male pelvis and a pelvic opening less than or equal to 1.0 indicates a female pelvis. In addition, generally the subpubic concavity is wider in females (Fp) than in males (Mp).

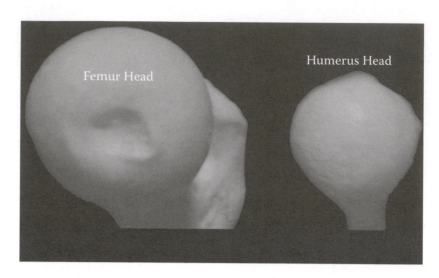

Figure 29.10a
The femur or thigh bone. The femur is the longest bone of the human skeleton. The human femur contains a longitudinal projection along its posterior (back) surface known as the "linea aspera," which is present for the attachment of muscle that is used for standing upright. It is only found in man, and thus is used in identification of skeletal remains. The linea aspera can also be used to determine whether the femur formed the left or right leg (right femur illustrated). The measurements depicted can be used to help determine the sex, age, and height of the individual.

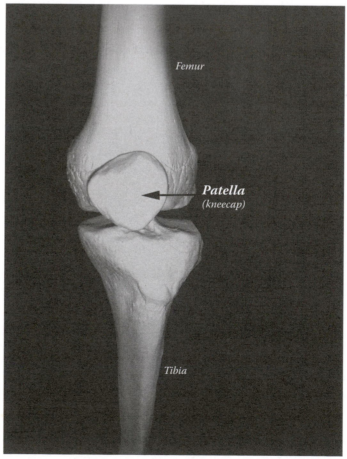
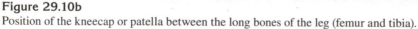

Figure 29.10b
Position of the kneecap or patella between the long bones of the leg (femur and tibia).

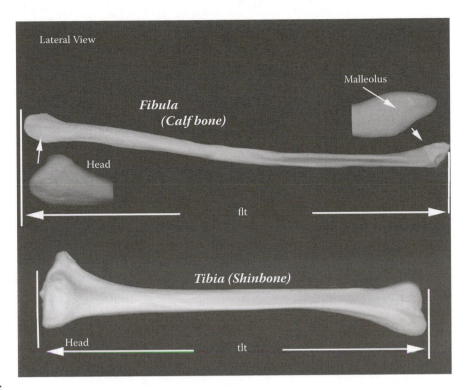

Figure 29.10c
The tibia (shin bone) and fibula (calf bone), the two long bones of the lower leg, showing some characteristic features. Their lengths can be used to help establish the age, sex, and height of an individual.

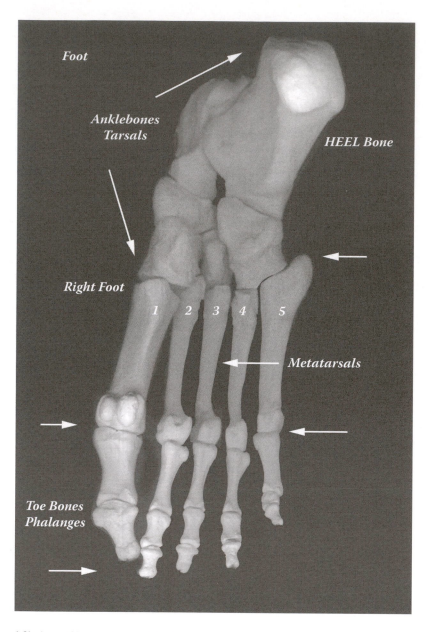

Figure 29.11
The tibia (shin bone) and fibula (calf bone), the two long bones of the lower leg. Their lengths can be used to help establish the age, sex, and height of an individual.

3. Draw and sketch the various skeletal elements provided to you by your instructor. Practice making linear and angular measurements on these skeletal bones. A comprehensive discussion of human osteology is given by W. M. Bass. The interested reader is referred to this text.

4. Prepare a data table and note the various linear and angular measurements you make of the known bones (see Table 29.1).

5. Calculate the cited indexes in Table 29.2 from the data in Table 29.1.

TABLE 29.1
Bone Data Table (the Included Bones are Suggestions)

Name of Bone	Length	Width	Angle	Circumference
Clavicle				
Scapular				
Humerus				
Ulna				
Radius				
Fibula				
Tibia				
Skull				
Mandible				
Patella				

TABLE 29.2
Index Table (Add Other Skeletal Indexes as Required)

Skeletal Index	Value
Maximum cranial index (MCI)	
Orbital index (OI)l	
Nasal index (NI)l	
Scapular index (SI)l	
Pelvis opening index (poI)l	

Glossary

Anterior — In front of or before.

Anthropology — The study of humans and their cultures.

Archeology — The study of past human civilizations.

Artifact — Any person-made object.

Carpus — The eight carpal bones of the wrist.

Cervical — The neck region.

Clavicle — Collarbone.

Coccyx — Tailbone.

Costal — Pertaining to a rib.

Cranium — The portion of the skull that encloses the brain.

Dentition — The type, number, and arrangement of teeth in the dental arch.

Digit — A finger or toe.

DNA — Deoxyribonucleic acid.

Entomology — The study of insects.

Femur — Thigh bone.

Fibula — The smaller of the two lower leg bones.

Humerus — Upper arm bone.

Hyoid bone — Horseshoe-shaped bone lying at the base of the tongue.

Ilium — One of the three bones that make up the hip bone.

Ischium — Lower portion of the hip bone.

Lumbar — The lower back between the thorax and pelvis.

Mandible — Lower jaw.

Maxilla — Upper jaw.

Medial — Toward the middle.

Medicolegal — Related to medical jurisprudence or forensic medicine.

Metacarpal — A bone of the hand.

Metatarsal — A bone of the forefoot.

Morphology — The science of structure and form without regard to function.

Nasal — Pertaining to the nose.

Occipital bone — Bone in the back part of the skull.

Odontologist — A dentist or dental surgeon.

Orbit — The cavity that contains the eyeball.

Osteology — The science of the structure and function of bone.

Osteomyelitis — Infection in bone.

Parietal bone — One of the two bones that together form the roof and sides of the skull.

Pathology — The study of the nature and causes of disease.

Pelvic opening — Upper pelvic entrance.

Pelvis — The bony structure formed by the hip bones, sacrum, and coccyx.

Phalanges — Finger or toe bones.

Posterior — Toward the rear.

Postmortem — After death.

Proximal — Nearest the point of attachment to the body.

Pubis — Anterior part of the hip bone.

Radius — The outer and shorter forearm bone.

Sacrum — The triangular bone at the base of the spine, normally made up of five fused vertebra.

Scapula — Shoulder blade.
Skull — The bony framework of the head, composed of 8 cranial bones, 14 facial bones, and the teeth.
Spine — The column of bones in the human back made up of 33 vertebrae.
Sternum — The breastbone.
Tarsal bones — The seven bones that make up the midfoot.
Temporal bone — A bone on both sides of the skull.
Thorax — The rib cage.
Tibia — The shin bone.
Ulna — The inner, longer forearm bone.
Vertebra — One of the 33 bony segments of the spinal column.
Zygomatic bone — Cheekbone.

30

Forensic Anthropology 2
Examination of Grave Site Bones

Teaching Goals

Each student will gain experience as a forensic anthropologist by examining, comparing, and identifying human bones.

Background Knowledge

Forensic anthropologists are frequently called upon by police investigators to examine, identify, and interpret human bones during death investigations. The forensic anthropologist is often asked: Are these bones human or another animal? What was the person's sex? How old was the person? What was the cause of death? The following exercise is designed to introduce the student to the basic methods of forensic anthropology.

Equipment and Supplies

1. Apparent skeletal remains from Experiment 28
2. Artifacts from Experiment 28
3. Soil specimens recovered in Experiment 28
4. Models of bones used in Experiment 29
5. Ruler
6. Micrometer
7. Protractor
8. Tape measure
9. Cleaning brushes
10. Specimen trays
11. Sieves
12. Graph paper

Hands-on Exercise

1. Using the knowledge, skills, and abilities you have acquired from Experiments 26 to 29, complete the following exercises.
2. Examine the skeletal remains acquired during Experiment 28. Examples of recovered questioned bones are given in Figure 30.1 through Figure 30.10. (Your instructor may provide you with different specimens.)

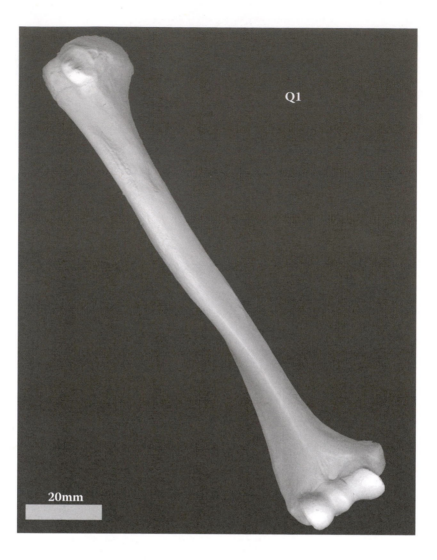

Figure 30.1
Q1 from grave site.

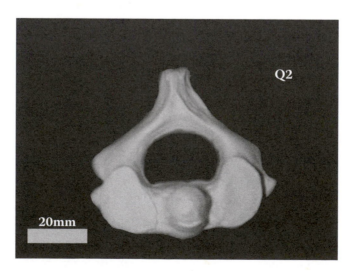

Figure 30.2
Q2 from grave site.

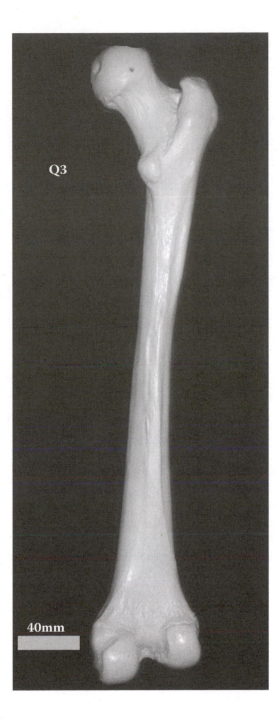

Figure 30.3
Q3 from grave site.

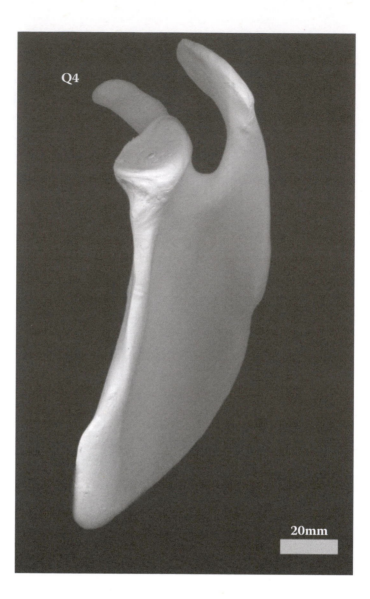

Figure 30.4
Q4 from grave site.

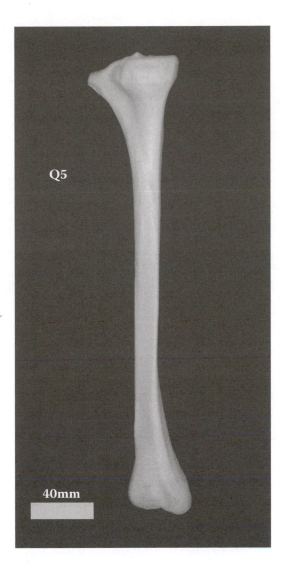

Figure 30.5
Q5 from grave site.

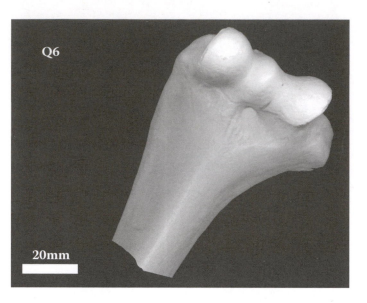

Figure 30.6
Q6 from grave site.

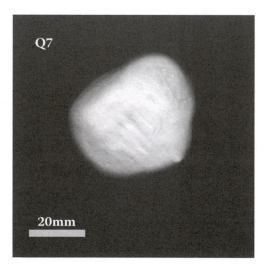

Figure 30.7
Q7 from grave site.

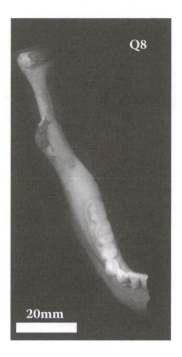

Figure 30.8
Q8 from grave site.

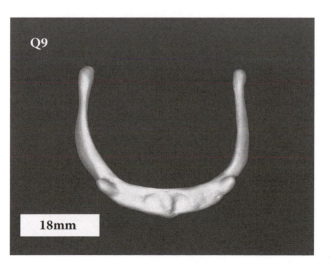

Figure 30.9

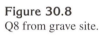

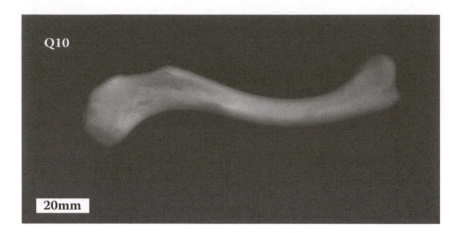

Figure 30.10

3. Examine each of the questioned specimens. Draw (on graph paper) and photograph (use a ruler) each questioned specimen.

4. Make the appropriate physical measurements (i.e., length, width, diameter, circumference).

5. Record the collected data in Table 30.1 and Table 30.2.

TABLE 30.1
Questioned Bone Data Table

Q-bone	Length	Width	Angle	Circumference	Identification

TABLE 30.2
Index Table

Skeletal Index	Value

Name _____   Date _____

Report

1. Answer the following questions for each questioned bone:
   - Is the bone from a human? If yes, why? If no, why not?
   - Is the bone from a child or adult? Why?
   - Is the bone from a male or female? Why?

2. Identify each of the questioned bones.

Q1 _____

Q2 _____

Q3 _____

Q4 _____

Q5 _____

Q6 _____

Q7 _____

Q8 _____

Q9 _____

Q10 _____

31

Digital Photography Image Processing

Teaching Goals

Each student will learn the basics of digital image processing.

Background Knowledge

Traditional photography mimicked the human eye by taking an image formed by the interaction of light with a subject and projecting an image of that subject with a lens onto a light-sensitive material. The ultimate purpose is to form a two-dimensional image of a three-dimensional natural subject that is recognizable as the original subject. Unlike traditional photography, digital imaging forms an image of a scene or subject by dividing the projected image into an extremely large number of tiny picture elements (known as pixels) organized into a mosaic array or pattern.

The primary advantage of digital photography over traditional photography is straightforward. Since digital images are composed of very large numbers of tiny digital elements, or pixels, made up of distinct sets of numbers or digits rather than fixed pictorial elements composed of different colored dyes embedded in layers of film, it is easy to use a digital computer to work with the images.

Equipment and Supplies

1. Digital camera (at least 3.0 megapixels) or digital scanner
2. A computer with Adobe Photoshop
3. Tripod
4. Various documents (i.e., checks, note pad)
5. Prepared written documents
6. Prepared altered documents
7. Scale or ruler

Hands-on Exercise

1. Scan or photograph the questioned specimen.
2. Import the image into Photoshop; save as a .tif file.
3. Open suspected written file (Figure 31.1a).
4. Process known checks 3401 to 3403 to show alteration. The known checks are processed using Adobe Photoshop (see Figure 31.1b–d).

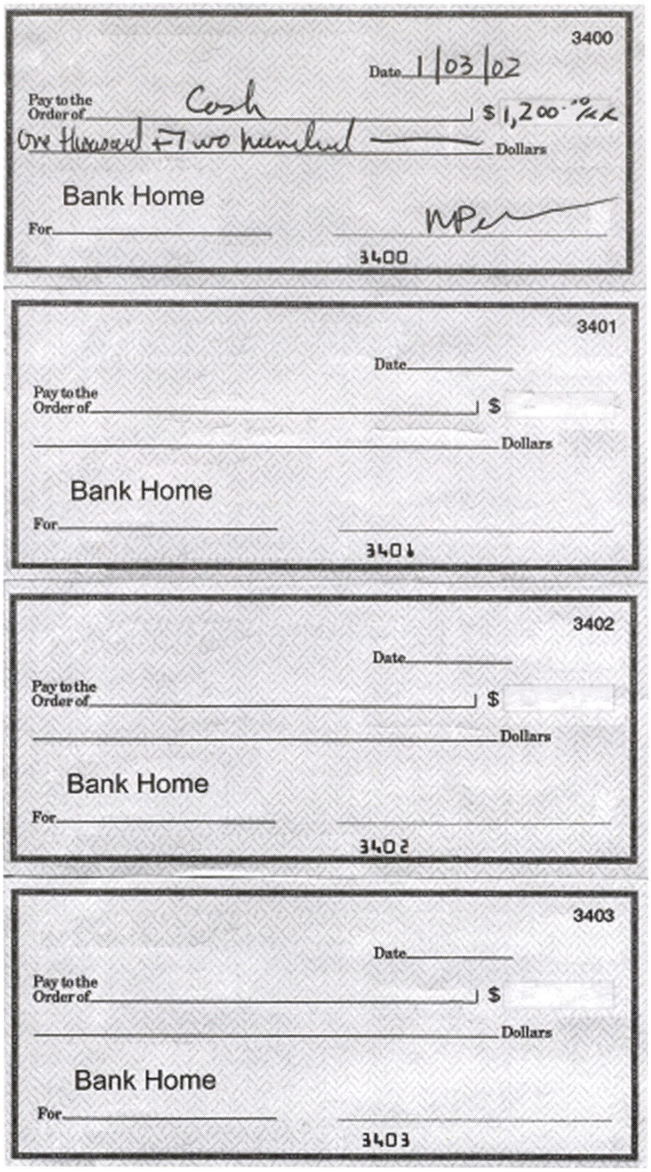

Figure 31.1a
Depicted are four checks. Check number 3400 is believed to have been altered in dollar amount. Checks 3401 to 3403 are known checks from the original checkbook.

5. The checks are processed as follows:
   a. Open the Image file on the Toolbar.
   b. Open the Adjust file and go to the "levels" option.
   c. Change the "levels" input value.
6. Focus in on one of the checks and enlarge (see Figure 31.1d). Note the check amount has been changed from $200.00 to $1,200.00.
7. Scan or photograph the altered document.
8. Import the image into Adobe Photoshop; save as a .tif file.
9. Open altered document's file (see Figure 31.2a).

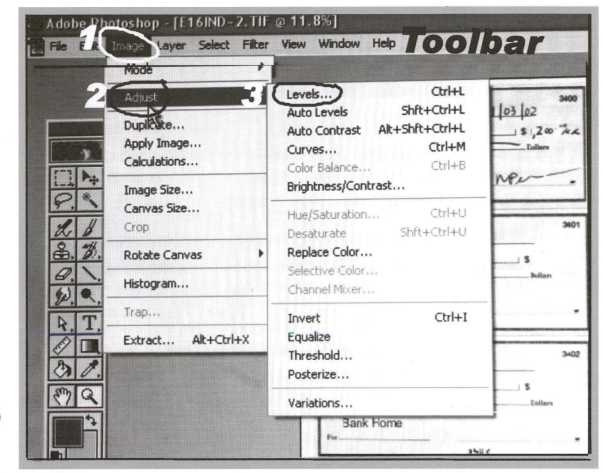

Figure 31.1b
Checks are processed as follows: (1) open the Image file on the Toolbar; (2) open the Adjust file and go to "levels."

10. Process the altered check as follows:
    a. Open the Image file on the Toolbar.
    b. Open the Adjust file and go to "curves."
11. Figure 31.2c shows the original alteration and its appearance after processing with "curves."
12. Finally, process the altered check with the "invert" and "levels" options combined (see Figure 31.3).
13. Try to detect the alterations with different menu options.
14. Make your own documents, alter them, and detect the alterations with your own combination of Adobe Photoshop menu options and applications.

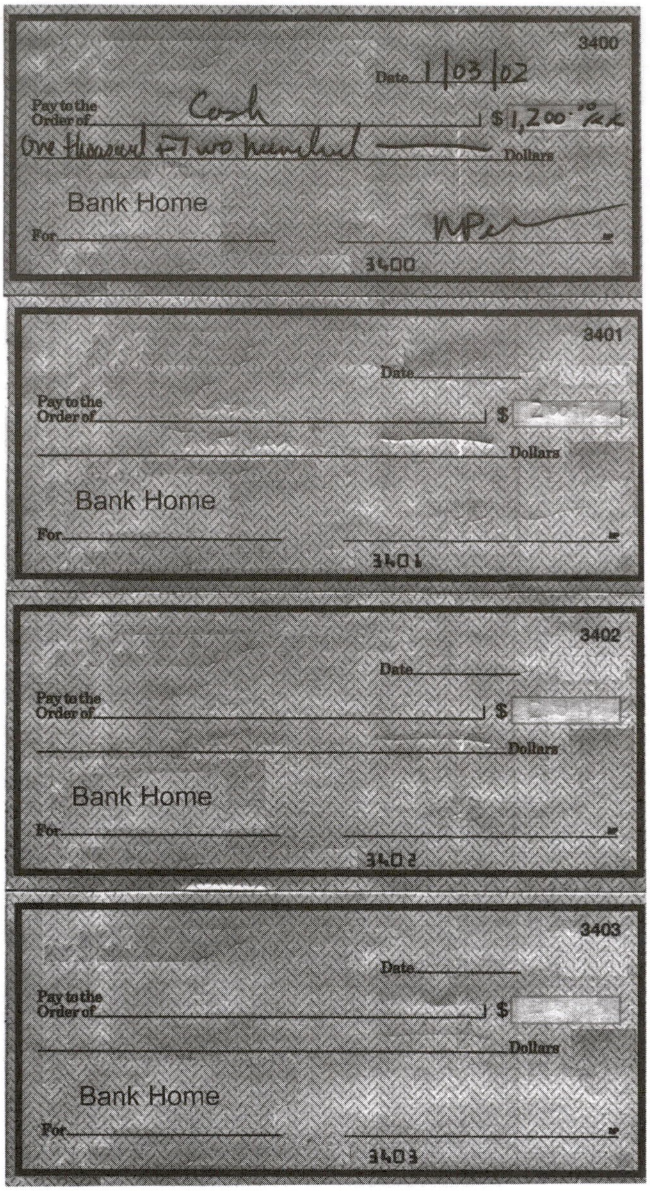

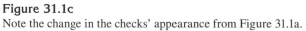

Figure 31.1c
Note the change in the checks' appearance from Figure 31.1a.

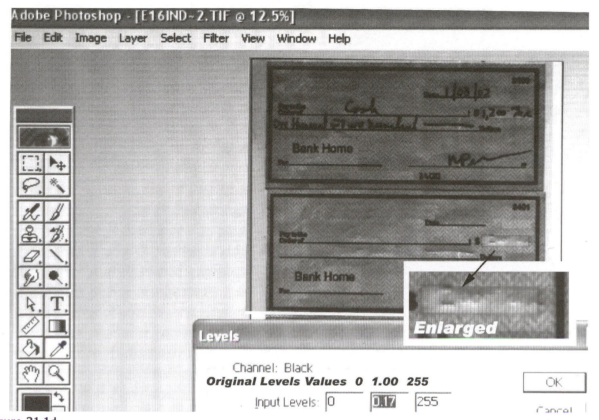

Figure 31.1d
Shows how changing the "levels" input value makes the alteration obvious from the appearance of the writing on the known checks.

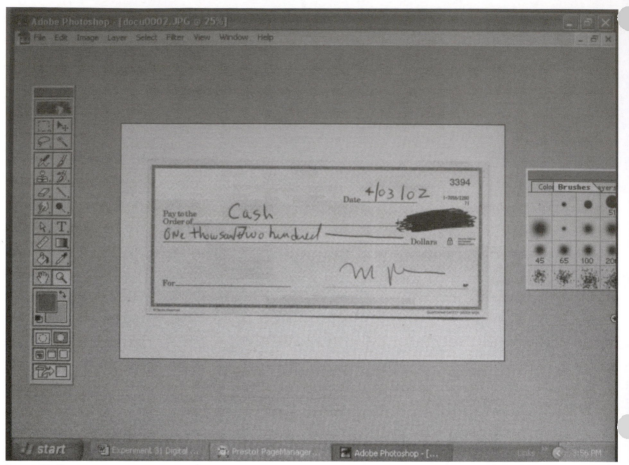

Figure 31.2a
Altered check.

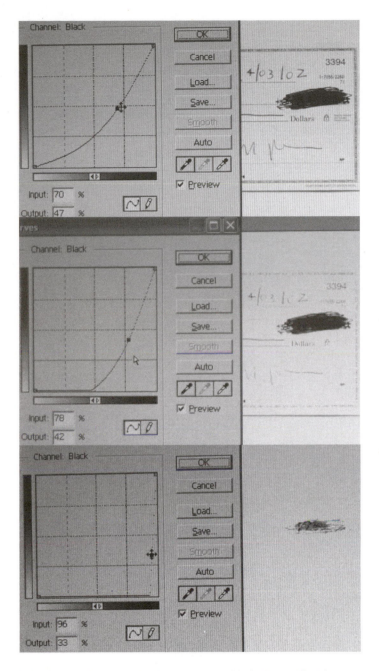

Figure 31.2b
The appearance of the altered check as it is being gradually processed with the "curves" option.

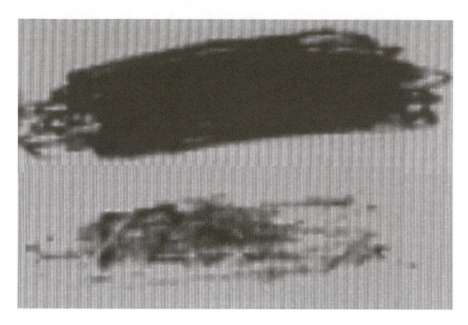

Figure 31.2c
(Top) An enlargement of the original alteration. (Bottom) The original alteration and its appearance after processing with "curves" (1200).

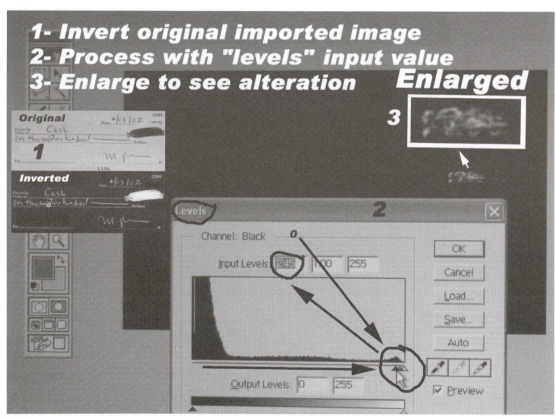

Figure 31.3
Processing the altered check with a combination of options.

Part 2

A number of the users of the first edition of this laboratory manual have requested that we include a small number of exercises that involve the use of the fundamental chemical instrumentation routinely found in college chemistry laboratories and available in some advanced high school programs. Each exercise has been designed to incorporate a lesson of forensic importance. They involve chromatographic separations and identification as well as the use of chromatography for qualitative analysis of volatile materials. Exercises in visible and ultraviolet electronic molecular spectroscopy are included that will elucidate the relationship between transmission and absorbance, as well as the relationship between absorbance and concentration. Molecular vibrational spectra will be introduced as infrared transmission measurements performed on polymer films.

Prior to each of the separate sections dealing with spectroscopy and chromatography there is a short discussion of the theory of each topic, in addition to drawings and diagrams that will assist the student in understanding the basic concepts involved. The individual exercises will then add to the general explanations with a short, specific narrative dealing with the particular experiment.

These experiments are optional and of an advanced nature for those who are not science majors. They are not required for those whose interest is in the nature of a survey of laboratory criminalistics. They are, however, designed to add to the advanced student's knowledge and understanding of chemical instrumental measurements in general and the application of these measurements to forensic science.

At the end of the manual prior to the bibliography, the students will find a collection of infrared spectra that are composed of data based polymers and food wrapping that have been analyzed and identified in our laboratory. Students may want to employ these in Experiment 33.

Experiment

Chromatography 2
Identification of a Single-Component Solvent by Gas Chromatography

Teaching Goals

The student will learn the operation of a gas chromatograph and will be able to identify each component of the instrument and should be able to describe its function.

Background Knowledge

Chromatography is a chemical and physical process that can be employed to separate the various components of a mixture into individual chemical compounds. Chromatography was originally described by Russian botanist Mikhail Zweitt, in the early 1900s, as a method of separating plant pigment mixtures such as chlorophylls and xanthophylls. He employed an active solid (finely divided calcium carbonate) packed into a rather large diameter glass column and a suitable solvent to elute the compounds of interest in different fractions of the added solvent. The name comes from the Greek *chroma* meaning color, and *graphein* meaning to write. The applications of chromatography have grown tremendously over the last century, primarily due to the development of several new types of techniques and the growing need of scientists to separate complex mixtures of chemical compounds.

The process involves variations in the relative attractiveness of different compounds (the ones to be separated) of similar chemical composition for a stationary chemical (stationary phase) as opposed to a moving fluid (mobile phase), which is called the eluent. The stationary phase is immiscible in the mobile phase. The stationary phase may be a solid composed of fine particles with a large surface area or it may be a liquid absorbed onto the solid of another surface, such as the inner walls of a capillary column. The mobile phase can be any fluid for which the analyte has a reasonable affinity or solubility. This is usually a gas (nitrogen, helium, or hydrogen) or an aqueous or nonaqueous liquid solvent system. Analytes that are strongly held to the stationary phase require a greater volume of the solvent to move them along or elute them, while those with less affinity require less volume.

A schematic of a gas chromatograph is shown in Figure 32.1. Here the columns are filled with an active stationary phase. The eluting gas, along with a mixture of components, is added to the top of the column at the injection port. As the gas passes through the column, the materials travel at different rates and finally each elutes separately from the column. A representation of the response of the output of an electronic detector to an eluting component is also shown.

Chromatographic methods can be classified in a number of ways: planer, column, liquid, ascending, descending, etc. based on certain parameters. When the mobile phase is a gas, the technique is called gas chromatography; when it is a liquid it is referred to as liquid chromatography. The active stationary phase

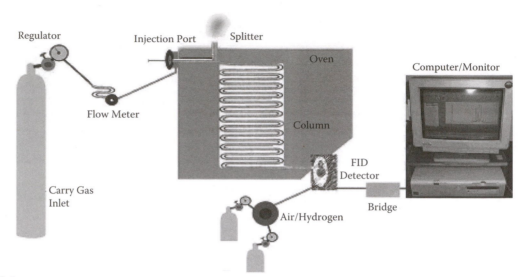

Figure 32.1
A diagram of a gas chromatograph.

can be contained in a column or be on a planer surface. The most common planer chromatographic method has the active stationary phase (absorbent) bound to a flat glass plate or polymer sheet. This technique is referred to as thin-layer chromatography (see Experiment 21).

In gas chromatography, when packed columns usually 1/4 in. or 1/8 in. in diameter and up to about 20 ft in length are used, usually the pressures employed to force the mobile gas phase through the column are no more than 150 psi. In many modern gas chromatographs, capillary columns are used that may have long lengths (30 m or more) and narrow bores (0.25 to 0.32 mm) and need only about 30 psi to maintain the required flow of the mobile phase. In these gas chromatographs, it is often necessary to limit the amount of sample introduced onto the column so that it is not overloaded. A technique called sample splitting is employed so that only a small portion of the analyte injected into the instrument is introduced onto the column, with the remainder vented into the atmosphere or trapped for disposal as waste (see Figure 32.2 and Figure 32.3 for images of a modern gas chromatograph and its data output with an electronic computing integrator).

In liquid chromatography, when high efficiency (the ability to separate a large number of similar compounds in a short time) is desired, columns measuring about 10 in. in length and 1/4 in. in diameter are used that are packed with very fine particles. In order to be able to force the mobile liquid phase through the tightly packed particles, pumps capable of generating pressures of several thousand pounds per square inch are employed and the technique is called high-performance liquid chromatography (HPLC) (see Figure 32.4).

In the classical method where dyes are separated by thin-layer chromatography or column methods, localization of the compounds of interest is easily accomplished by visual examination. When the compounds are not visible or are in the vapor state, as in gas chromatography, a specialized device capable of generating an electrical signal when they are present is required. We refer to these as detectors, and they are based on a plethora of the chemical and physical characteristics of the compounds of interest. Some of these are destructive in nature, in that the compound detected cannot be recovered, while others are not destructive, even though the samples actually tested are often not recovered. For a thorough discussion of these topics, the interested reader is referred to any comprehensive text in chemical instrument analysis.

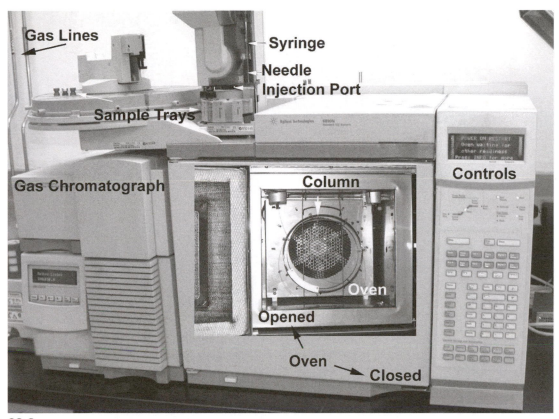

Figure 32.2
A gas chromatograph, showing the injection port, oven, column, and controls.

In gas chromatography and liquid chromatography, the temperature and composition, respectively, of the mobile phase must be controlled to guarantee the reproducibility of the retention data. Temperature control is very important in gas chromatography, and it can be ramped or programmed to increase with time as a method of solving difficult separation problems. In HPLC, the solvent composition of the mobile phase can be altered (gradient elution) during the analysis to accomplish a similar goal to that of temperature programming in gas chromatography.

All of the methods mentioned above, and others, can be employed to separate complex mixtures. The instrument methods can also be readily employed to qualitatively identify and quantify the components. These instrument methods rely, as mentioned above, on a detector to sense when a component is eluting from the instrument. This detector outputs a constant signal (usually a voltage in the tens to thousands of millivolts range) that changes when a component is detected. The signal is monitored over time, most simply with a strip chart recorder, but more commonly today with an electronic integrator or computer data system, as seen in Figures 32.3 and 32.5. These latter two devices provide a readout of the data, including retention times, peak heights and areas, and other information, often in conjunction with a graphic plot of the data that is called a chromatogram.

Qualitative identifications are rudimentally based on the raw retention time. Adjusted retention times are a better value to employ. For complex chromatograms, referencing the adjusted retention time to the adjusted retention time of an internal (added) standard gives better reproducibility of the retention data over extended periods. This term is called the relative retention. For the best transferability of data, the retention index method is best.

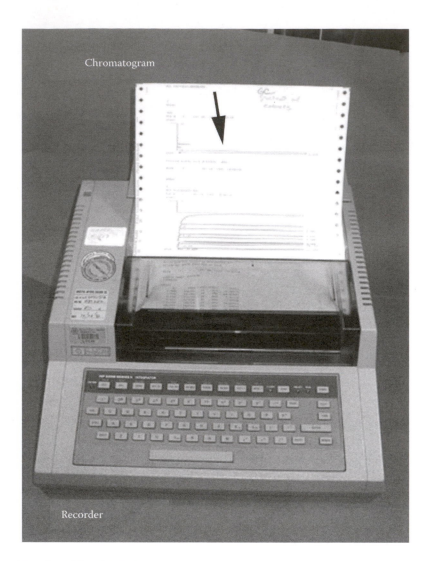

Figure 32.3
A computing integrator/recorder with a chromatogram.

For quantitative analysis, the size of the peaks is important. The total integrated peak area is proportional to the amount of material eluted from the column. If the peak is symmetrical and not overly wide, the peak height is a good estimate of area, and therefore of the amount eluted. In gas chromatography, many of the peaks lack the symmetry to allow peak heights to be employed for accurate quantitative results, therefore measurement of peak area is the preferred method. The peaks generated in HPLC tend to be sharp and symmetrical, allowing peak heights to be used routinely for quantitative analysis.

Because the actual response of the various detectors employed in either HPLC or gas chromatography varies with the actual compound being detected and measured, calibration is necessary. That is, because the signal generated from equal amounts of various chemicals will differ, it is necessary to construct a calibration curve or determine a calibration (response) factor for the material whose amount is to be determined. The two methods most often employed for calibration are the external and internal standard methods.

In the external method, different amounts of the analyte are injected into the instrument and the differing responses plotted against these amounts generate an analytical calibration curve. This method requires highly reproducible injection volumes of the different calibration solutions to be made. This cannot routinely

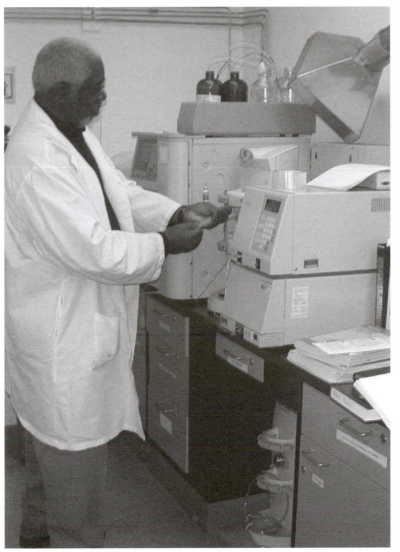

Figure 32.4
HPLC being used to separate and detect mixtures of explosives.

be accomplished with the syringe technique that is employed in gas chromatography because it is very difficult to load, measure, and inject the equal amounts necessary. The method employed for introducing samples into a high-performance liquid chromatograph is a sampling valve equipped with a sample loop of known volume. Use of this technique results in the introduction of reproducible volumes of calibration solution, which is amenable for utilizing the external calibration method and is widely employed for HPLC.

In the internal standard method, amounts of a compound different from the analyte are added to all of the analyte calibration solutions so that the concentration of this internal standard is the same in each. The ratios of the peak areas of the analyte to those of the internal standard are determined and these are plotted against the concentration of the analyte in the calibration solutions. The resulting calibration curve can then be employed to determine the concentration of an unknown if the same internal standard is added to the unknown at the same concentration as that used for the calibration curve solutions. This method corrects for the inability to inject constant volumes with a syringe and for variations in the injection technique that lead o alterations in the peak shapes of the analytes.

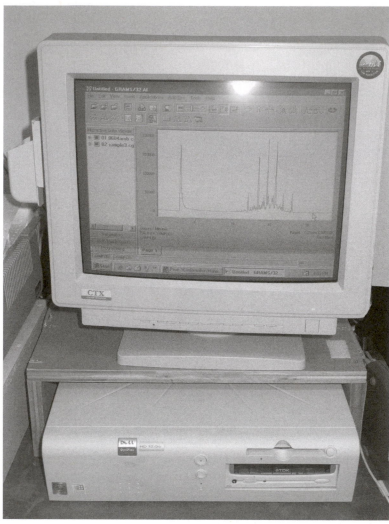

Figure 32.5
A computer monitor showing a chromatogram of an arson residue analysis.

Various retention data can be collected. The easiest one to determine is the raw retention time, which is the time from injection to the maximum of the eluted peak. A single component can be correctly identified if the population of possibilities is a closed set and all the retention times for these materials are different. For a general unknown, or for a mixture of materials, identification can be more problematic. Multiple columns of different polarities of the stationary phase can be of great assistance with this problem. The use of methods such as gas chromatography/mass spectroscopy or gas chromatography/infrared spectroscopy is more conducive to this type of analysis.

Discussion of these topics is beyond the scope of this manual and the interested reader is referred to the literature.

There are two methods of temperature control for the chromatograph oven. The first is isothermal, which means that the experiment is conducted at one temperature. In the second, temperature programming, the temperature is increased while the experimental separation is in progress. Temperature programming helps solve the classic separation problem, but it will not be necessary for this exercise.

Equipment and Supplies

1. The unknown will be one or a mixture of no more than three of the following: normal alkanes, heptane (C_7), octane (C_8), decane (C_{10}), cyclohexane, toluene, acetone, and 1- or 2-propyl alcohol. Separate containers containing each of these as references will be required.

2. A gas chromatograph equipped with either a packed or capillary column capable of performing the separation of the above solvents when operating isothermally will be required. The detector may be either a thermal conductivity or a flame ionization type, with the latter preferred. Nitrogen or helium may be the carrier gas, with helium preferred for the thermal conductivity detector. A strip chart recorder or electronic data system will be required for detector output.

3. Various pieces of glassware, microsyringes (10-μL capacities work well), a wash solvent like ethyl alcohol that can be used to remove organic solvents from the syringe, followed by a water rinse.

Hands-on Exercise

1. The instructor will set the experimental conditions on the gas chromatograph, such as column oven temperature, injector temperature, detector temperature, and gas flow rates (see Figure 32.6). The instructor will also set or instruct you on how to set the proper parameters for the output devices.

2. Obtain each reference standard. After injection (the teacher will supply the recommended injection volume) into the chromatograph, determine the retention time for each material. The best procedure is to make at least three injections of each and find the average for your retention data. These will be used to identify your unknown's composition. You will notice that under isothermal conditions (oven set to a fixed temperature), the later the

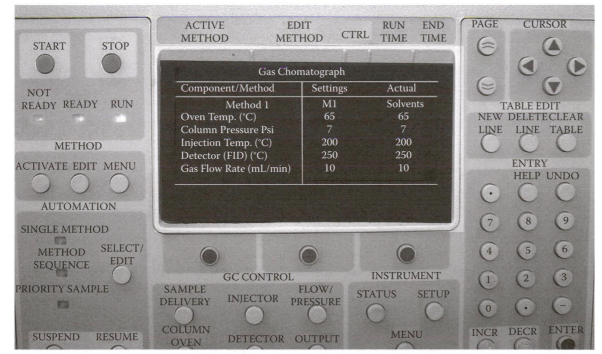

Figure 32.6

Typical parameters used in a gas chromatography isothermal run to separate familiar organic solvents (i.e., methanol, ethanol, propanol, acetone, pentane, hexane, heptane, and so on). The oven temperature is held constant for 10 minutes to complete each run.

peak elutes, the wider it tends to be, and if the same concentrations are employed for each, the relative height of the peaks will be lower for the later eluting materials. Record the retention times and their averages for each material in your notebook.

3. Obtain an unknown from the instructor. After injection into the chromatograph (three replicates should be performed), determine the identity of the solvent present in the sample. Note that there may be more than one component in the sample. Record in your notebook the retention data for the components in this mixture and the calculated average. Record your identifications.

Optional Exercise

The instructor may be able to set a temperature program on the gas chromatograph for your evaluation. You know the relative retention time for the solvents (which one elutes first, the order, and which one is last).

1. The instructor will supply a solvent sample that is a mixture of the reference standards above.

2. Analyze the sample with the isothermal conditions employed above.

3. Employing the temperature program set by the instructor, make an injection of the mixture and obtain a chromatogram of the sample.

4. Compare your results to those obtained by the isothermal method.

Name _____ Date _____

Report

1. Report the average retention times for each of the known solvents.

2. Report the average retention times for each component in your unknown sample.

3. Report the identity of each of the components in your unknown sample.

4. If you have a photocopy machine available, supply copies of your chromatograms along with your report.

Spectroscopy 1

Methods for the Identification of Materials Based on Their Absorption of Light of Various Wavelengths and Identification of Polymer Films by FTIR

Teaching Goals

Each student will learn to operate a basic infrared spectrophotometer. Each student will examine a number of polymer films obtained from food and cigarette packages with this instrument. In addition, each student will learn how to identify the class of polymer.

Background Knowledge

Light is that portion of the range of wavelengths of electromagnetic radiation from 10 to 400,000 nanometers (1 nm = 10^{-9} m). The portion of this range between 10 and 390 nm is referred to as ultraviolet (UV) light, the portion between 390 and 780 nm is visible light, and the portion from 780 to 400,000 nm is infrared (IR) light.

Electromagnetic radiation can be thought of as packets of particles of energy called photons traveling in an oscillating manner in their direction of propagation so they act as a sine wave. This is a very simple explanation that can be employed to explain what is referred to as the dual nature of light; that is, it behaves as a wave or a particle depending upon the experiment one performs. The nature of this behavior is too complex to discuss in this manual. Students with additional interest are directed to textbooks on physics or chemistry, as well as treatises on the nature of light.

Light has the following characteristics: Wavelength (λ) is the distance between repeating portions of the sine wave. Velocity (c) is the speed of light, which is a constant for all wavelengths in a vacuum. The amplitude (height of the wave) is related to the square of the intensity. The velocity of light in any other medium varies with the refractive index (RI) of that medium and the particular wavelength. The variation of the RI with wavelength is called optical dispersion. The frequency (ν) with which the light wave vibrates (cycles from maximum amplitude to its minimum) is measured in cycles per second, or hertz (Hz). The vibration direction of the light is perpendicular to its propagation direction. Normal light vibrates in all directions (360°). Light can become polarized, that is, the vibration direction can become preferential. When there is one preferred vibration direction as the light travels, it is said to be plane polarized. The energy of a particular type of light is directly proportional to its frequency and inversely proportional to its wavelength. The frequency and wavelength are related by Equation 1. What we perceive visually as color is actually related to the frequency, but is often referred to as being a function of wavelength; as the wavelength increases, the frequency decreases and the color shifts toward red, with the energy of the light decreasing. Conversely, the shorter the wavelength and the higher the frequency, the higher the energy.

$$\text{Frequency } (\nu) = \text{Velocity } (c)/\text{Wavelength } (\lambda) \qquad (1)$$

There are a large number of interactions that light undergoes with matter. Reflection, refraction, scattering, change in velocity, gain or loss of energy, and loss of intensity are some of these interactions. For the purposes of the exercises in spectroscopy in this manual, we will concentrate on the absorption of light as a function of the wavelength. This absorption causes a loss of intensity from the initial beam of light interacting with the sample. This loss of intensity is measured as a ratio of the intensity of the light measured (I) when the analyte is present to the intensity of the light when everything is present in the sample except the analyte (I_0), or I/I_0. This is called the transmission (T), and when multiplied by 100 it is the percent transmission. This is done by measuring the transmission of a reference sample (I_0) followed by that of the sample (I) when single-beam instruments are employed. In the more sophisticated double-beam instruments, both the reference and sample transmission are measured together and the readout of the ratio is in percent transmission.

If the percent transmission is measured as a function of the wavelength, or other units, and the data are plotted, the result is called a spectrum. The location of the maximum or maxima with their ratios, the minimum or minima, and the shape of the spectra can be employed to identify the compound and the size of the peak maxima can be used to quantify the analyte. Unfortunately, the size of the transmission peak is not related linearly by a straight line to the concentration or amount of analyte present. A number of researchers (Beer, Lambert, Bouguer) have investigated the various measurement conditions that affect the percent transmission recorded. They discovered that the path length of the light beam through the sample, the concentration of the analyte, and a characteristic constant for the particular compound at a specific wavelength, called the extinction, were the most important controlling factors. The combined relationships in Equation 2, most often referred to as "Beer's law," present a relationship where the concentration of the analyte and a term called the absorbance (A) are a straight line. This equation also reveals the relationship between absorbance and transmission.

$$A = abc = -\log T = \log 1/T, \qquad (2)$$

where A is the absorbance, b is the path length (kept constant during a series of measurements), c is the concentration of the analyte (usually in moles per liter, but sometimes as an amount of analyte as a weight in a constant volume of solvent), and a is the absorptivity, which is a proportionality constant and has the inverse dimensions of the other factors (b and c) so that A is a dimensionless number. Absorbance and absorptivity are related for this relationship at a specific wavelength. Today, especially in the United States, absorptivity is usually employed as the molar absorptivity and the concentrations are in moles per liter.

The spectroscopy with which we are concerned in this exercise is called molecular absorption spectroscopy, as we are concerned with absorption of light by molecules which causes the energy content of the molecule to increase. This increase in the energetic state is called an excited state, and in most cases it lasts for only a very short time, usually less than 10^{-8} sec. Depending upon the frequency (energy) of the photons, this excitation is either of the electrons in the molecule or the vibrational energy between the atoms bound together to form the molecule. These excited states are called excited electronic and vibrational states, respectively.

When the absorption of UV or visible light is employed to generate a spectrum it is called molecular electronic absorption spectroscopy. When atoms are bound together by either transfer or sharing of electrons, a bond or multiple bonds are formed and these electrons are referred to as valence electrons. These bonding electrons are held in molecular orbits so that the electron energy of the molecule is lower than that

)f the sum of the individual atoms. These bonding orbitals are called sigma or pi bonds. Electrons that do not enter into bonding are called nonbonding electrons. Orbitals that have higher energy than the atoms are called antibonding orbitals, and if they contain electrons, result in a lessening of the total strength of the bonding and are designated sigma* and pi*. When photons of the correct energy, equal to the difference between the bonding or nonbonding orbitals and the antibonding orbitals, impinge upon the compound, some of them will be absorbed, momentarily promoting the electrons from the lower energy level to the antibonding state. This results in the compound having a higher total energy than normal for a very short period of time until it relaxes to the ground state. This absorption is the reason for the loss of intensity of the initial light beam that we measure. The energy of the transitions from nonbonding to pi*, pi to sigma* are often in the visible and UV light range.

These transitions have reasonably strong absorptivities so that dilute solutions (0.01 molar or less) can be qualitatively analyzed and quantitation of low actual concentrations can be performed. More concentrated solutions are determined by dilution of the sample. It is these absorptions, when they take place in the visual region, that result in our perception of color. The color we observe is the complementary color of the light that is absorbed. This means that the color seen is white minus the color absorbed. Molecules that have an electronic configuration such that they absorb well in the visible or UV region are said to have chromophores. Because these chromophores can occur on a multitude of molecules, their spectra are not specific to a particular chemical. If other factors are determined such as the absorptivity or shape of the curve and derivative spectra are plotted, the specificity of an identification based on UV–visible spectroscopy can be significantly increased to near specificity.

When light in the infrared region of the electromagnetic radiation spectrum is absorbed and the transmission is determined as a function of wavelength or frequency, this is called molecular vibrational spectroscopy. There are a number of ways that vibrational spectra can be determined, but we shall only deal with those obtained by the infrared transmission technique. The display of the spectra as transmission or absorbance versus wavelength, frequency, or wave numbers per centimeter is not critical for understanding of processes involved or the usefulness of the technique. Today the x-axis is most commonly displayed as cm^{-1}, while the y-axis is plotted as either transmission or absorbance with equal frequency. Absorbance is becoming more popular because computer searchable databases are usually stored as absorbance data. Similar to UV–visible light spectroscopy, when a photon of the proper energy impinges upon a molecule it is absorbed if the differences in energy between the ground state and an excited vibrational state are equal to the photon's energy. These energies are in the infrared region of the spectrum and describe the energy of vibration of atoms that are held together by single or multiple bonds. The spectra obtained are quite complex and in most cases can be considered a specific identifier for the molecule. Tables of group frequencies have been published where the vibrational energies of various functional groups have been assembled. This allows the analyst to make determinations about the groups that are present in a molecule even if he does not have reference spectra for comparison.

Spectrophotometers are similar in many respects, but differ in the details of the hardware and the method by which a spectrum is generated. Instruments are either single beam or double beam. In single-beam instruments, the reference and sample spectra are determined independently and the ratio is calculated either manually or automatically by a computer. In double-beam devices, the ratio of the sample's transmission to the reference's transmission is determined simultaneously in real time.

A schematic of a classically designed simple spectrophotometer is shown in Figure 33.1. Here there is a source, sample holder, wavelength selection device, detector, readout device, and all the necessary optics to control the light beam. For a wavelength selector in the classic design, a monochromator is used with a light-dispersing device such as a prism or diffraction grating, with the necessary wavelengths selected by the exit slit.

Recently single-beam instruments employing linear diode array (LDA) detectors have become commonplace for routine UV–visible light determinations. These units have dispersing elements, but no

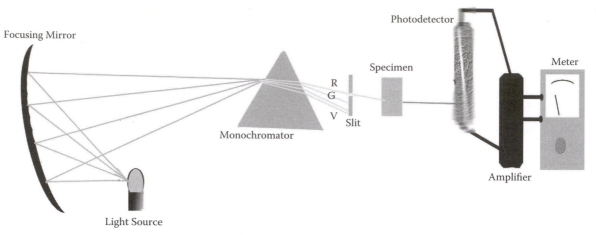

Figure 33.1
Schematic of a spectrophotometer.

monochromators, and utilize computers to control the detectors, calculate the transmission or absorbance, and display the spectrum.

In the last 15 years, as computers with faster processors and greater storage capacity have become widely available, infrared spectrophotometers based on Michelson or other interferometers, wherein the interaction of all wavelengths with the sample is recorded simultaneously as an interferogram, have become popular. This interferogram, which represents a time domain spectrum, can be rapidly converted to a frequency domain spectrum by a mathematical transformation know as a Fourier transform. The transformation is easy to perform with modern computers. Figure 33.2 shows a Fourier transform infrared spectrophotometer (FTIR).

Figure 33.2
A typical Fourier transform infrared spectrophotometer. Courtesy of John Jay College.

Equipment and Supplies

1. Your instructor will explain the type of spectrometer available to you and its principle of operation. There are a number of excellent references that can be consulted for detailed explanations of the theory and operation of these instruments. In addition, your instructor will give you detailed instructions on how to operate the instrument in your laboratory. Pay attention and follow the instructions, as failure to do so may damage the device.

2. A number of plastic (polymer) films obtained from cigarette and food packaging will be used. If the film has been in direct contact with the food product it should be washed carefully to remove any contamination. Use soap and water followed by a thorough water rinse, then wipe with an alcohol swab and set aside to dry completely.

3. Clean scissors or scalpel (razor) blade and forceps.

4. Commercial polymer film holders designed for the specific instrument to be employed. If not utilizing a scalpel blade, an adequate holder can be constructed from thin cardboard and cellophane tape.

5. Use of a photocopy machine (if available) to copy the spectra obtained for the unknowns.

Hands-on Exercise

1. Follow the directions of the instructor concerning operation of the instrument. Instruments by different suppliers have different operating procedures that need to be followed in order to obtain correct results.

2. Determine the proper operation of the instrument by collecting a reference spectrum on the polystyrene film supplied by the manufacturer. Plot a hard copy and compare your spectra with the reference data.

3. Mount at least five different films, one at a time, in the film holder and determine their spectra. Plot each spectrum.

4. Compare each of your spectra with those in the Reference section at the end of this manual and report your results.

Name _____ Date _____

Report

1. Comment on the correctness of the check (polystyrene) spectrum generated by you compared to the reference.

2. Identify the sample name(s) for each unknown material the instructor supplies for you to identify followed by your choice of the polymer(s) that best matches its spectra. Remember your unknown sample(s) may be a multilayered film. Compare to spectra 1 to 29 in Appendix 1.

3. Comment on the similarities and differences you note by comparing the spectra you generated.

4. If you have a photocopy machine available, supply copies of your spectra along with this report.

Experiment

Spectroscopy 2
Use of Visible Spectroscopy in Color Determination

Teaching Goals

Each student will learn the basic operation of a visible spectrophotometer. In addition, each student will learn how to compare the spectra generated and recognize the similarities and differences between various molecules.

Background Knowledge

The student is directed to read and understand the portion of this manual that introduces the student to spectrophotometry (Background Knowledge in Experiment 33), particularly that on ultraviolet (UV)–visible spectroscopy. The various chromophores that are present on a molecule can often be determined from the absorption band so that a scientist can establish some information about the structure of materials. Examples of some common chromophores are shown in Figure 34.1. The easiest way to identify a chromophore is by comparison to a database of known wavelengths for specific chromophores. Dyes, particularly solutions, are readily examined by visible spectroscopy using the transmission method. Even mixtures of dyes can often be determined by careful examination of the questioned spectra and the reference database. In many cases the dye mixtures can be quantitated to determine the concentration of each in the solution. This multicomponent quantitative analysis is beyond this introductory experiment; however, interested students are referred to the literature.

Equipment and Supplies

1. A visible spectrophotometer. It may be a normal dispersive or diode array spectrophotometer. Your instructor will explain the type available and its principle of operation. There are a number of excellent references that can be consulted for detailed explanations of the theory and operation of these instruments. In addition, your instructor will give you detailed instructions on how to operate the instrument in your laboratory. Pay attention and follow the instructions, as failure to do so may damage the device.

2. Deionized water, volumetric flasks, beakers, pipettes, balance, weighing paper, spatula, cuvettes for the specific instrument, and standard dyes listed below to prepare solutions.

3. Cuvettes (sample cells) used for containing materials to be measured can be composed of different materials that have different transmission characteristics. Make sure you are employing the proper type for your

283

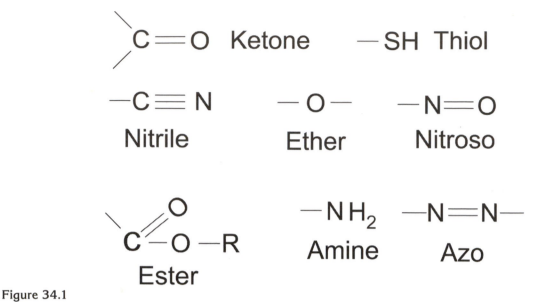

Figure 34.1
Some common chromophores.

measurements. Cells can be made of glass, plastic, fused quartz, and fused silica. **NOTE: Be careful of solvent compatibilities.** For visible spectroscopy, quality glass cells will be adequate.

4. Solutions of quinoline yellow (13 mg/L), erythrosine (red) (8 mg/L), indigotine (blue) (18 mg/L), each made in deionized or distilled water. These may be supplied to you prepared by the teacher.

5. Use a photocopy machine (if available) to copy the spectra obtained of the unknowns.

Hands-on Exercise

1. Follow the instructions of your instructor for the operation of the spectrophotometer in your laboratory and plotting of the spectra obtained.

2. If available, collect the spectrum from a reference material such as a holmium oxide filter to check the wavelength accuracy of your instrument.

3. Determine the spectrum of each dye solution, plotting absorbance versus wavelength from 350 to 700 nm.

4. Plot each spectrum.

5. Record the color observed, the maximum absorbance, and the color of the absorbed light in your notebook.

Optional Exercise

1. Choose one of the dye spectra and from the maximum absorbance calculate the absorptivity term "a" from the Beer equation, $A = abc$, assuming that the concentration "c" is in mg/L, the path length "b" is 1 cm, and A is in absorbance units.

Name _____ Date _____

Report

1. Comment on the wavelength accuracy of your instrument if the holmium filter was available. Record the reference wavelength and the difference between it and your determination for three to five peaks.

2. Report the maximum absorbance wavelength for each dye measured.

3. What is the color of the maximum absorbance wavelength? What is the color you observe visually? Report fo each dye measured.

4. If you have a photocopy machine available, supply copies of your spectra along with this report.

5. Comment in your notebook about your observations.

Quantitative Ethyl Alcohol Determination by Gas Chromatography

Teaching Goals

The student will be introduced to the operation and application of a gas chromatograph for the separation and quantitative analysis of ethyl alcohol. The student will be able to identify each of the instrument's parts and will be able to describe its function. In addition, each student will receive hands-on experience in making serial dilutions of standards and preparing standard calibration curves.

Background Knowledge

There are various methods for determining the area of the peaks generated. These depend on the output device available. The simplest is the use of a strip chart recorder. The area of the peaks can be determined by carefully cutting out with scissors the peaks for both knowns and unknowns. These are then carefully weighed on a balance and the ratio of the weights is used to determine the area ratio of the peaks. The absolute areas can be determined by weighing a known area of the same paper and determining the area by ratio calculation.

The peak areas can also be reasonably determined by constructing a straight baseline and tangents to each side of the peak using a straight edge. The width and height of the peak are then determined with a millimeter scale. The height times the width at 0.5 max (h × whm) is a good estimate of the peak area.

The best method for determining peak area is to use an electronic integrator or computer software integration system. The signal from the instrument to the readout device is in millivolts (peak height) and the length of the signal is in time (seconds). Therefore the signal integration of the chromatographic peak is in millivolt-seconds (mv-sec).

For the quantitative determination of ethyl alcohol, a calibration curve is constructed by plotting the ratio of the ethyl alcohol peak area over the n-propyl alcohol interval standard peak area against the ethyl alcohol concentration. The best-fit straight calibration line can then be determined by visual graphical construction or by manual calculation with a scientific calculator, calculating the best fit by means of least squares, or by employing an electronic spreadsheet. The concentration of the unknown is calculated by determining the ratio of the unknown ethyl alcohol peak area to the internal standard peak area either graphically or by applying the regression equation.

Equipment and Supplies

1. For this experiment, a gas chromatograph with either a special packed column or a polar capillary column capable of performing separation of the alcohols when operating isothermally will be used. The detector may be either a thermal conductivity or flame ionization type, with the latter preferred. Nitrogen or helium may be the carrier gas, with helium preferred for the thermal conductivity detector. A strip chart recorder or electronic data system will be required for detector output.

2. Various glassware, microsyringes (10 μL capacities work well), volumetric flasks, pipettes, and analytical balance. The wash and dilution solvent can be either distilled or quality deionized water.

3. A photocopy machine, if available, for copying your chromatograms.

Hands-on Exercise

1. The instructor will set the experimental conditions on the gas chromatograph, such as column oven temperature, injector temperature, detector temperature, and gas flow rate. The instructor will also set or instruct you on how to set the proper parameters for the output devices. In addition, he will explain how to determine the area of the eluted peaks.

2. Prepare the water solution of the internal standard to about 0.2 to 0.3% n-propyl alcohol by weight. Prepare standard solutions of ethyl alcohol in the range of about 0.1 to 0.5% by weight in water. These must be known accurately. Four solutions are sufficient, but five would be better. The more standards, the better the standard curve. Mix equal amounts of the internal standard solution with each of the calibration standard solutions; 0.5 or 1 mL of each should be sufficient. If the teacher does not supply a stock solution for dilution, prepare the stock, 1.0% ethyl alcohol solution, by weighing the proper amount of ethyl alcohol into a tared beaker that has water in it. Prepare the other standard alcohol solutions by serial dilution of the stock solution. Carefully label all containers. Your teacher will give you additional detailed instructions if you need to prepare the stock solution or other standards by dilution.

3. Prepare a standard curve by injecting about 0.5 μL of the standard solutions of ethyl alcohol you prepared in step 2, each of which has been mixed with an equal amount of the internal standard, into the gas chromatograph. Perform as many replicate injections as you feel are necessary to obtain reproducible results. Normally three are sufficient. A relative standard deviation of 10% or less is usually acceptable. Plot the area of ethyl alcohol standard/area of internal standard vs. the percent ethanol in each standard.

4. Prepare your unknown by diluting it with internal standard as in step 3 for constructing of the calibration curve. Make a series of injections and determine the average results. Use the calibration curve to determine the percent ethanol in your unknown. Report the value that was determined by the peak areas.

Optional Exercise

The instructor may assign 1, 2, or 3 below or a combination of them as optional exercises.

1. Using the data collected above for the standard curve, plot peak height ethyl alcohol standard/peak height internal standard vs. percent ethyl in the standard. Then determine and report the percent alcohol in your unknown from this calibration curve using the peak heights of the unknown injections.

2. Prepare a water solution of methanol and isopropyl alcohol at about 0.2% by weight (these can be together in one container). Mix equal amounts of an ethyl alcohol standard of about 0.3% by weight and the internal standard. Inject the usual amount into the gas chromatograph. You have already determined the retention time for the ethyl and n-propyl alcohol. You should now be able to observe that the chromatograph separates the four alcohols. Keep in mind that the first alcohol peak is methyl alcohol and the unidentified peak is isopropyl alcohol. Keep in mind that often the first peak eluted may be a water peak, and if a thermal conductivity detector is used, it may be an air peak followed by water. Speak to your instructor if you have questions.

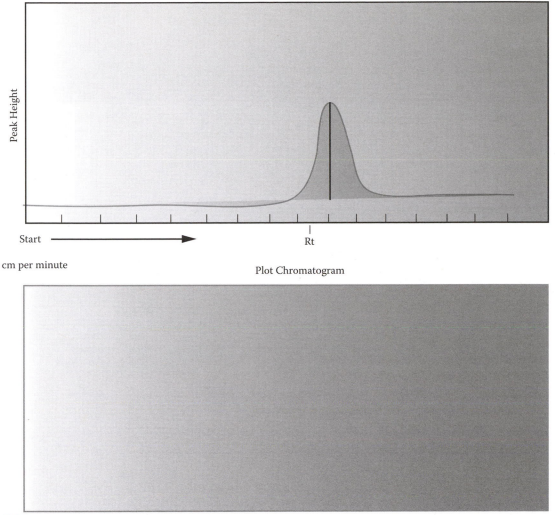

Peak Height

Start ———————————————▶   Rt

cm per minute   Plot Chromatogram

Figure 35.1

3. Your instructor may require you to calculate the relative standard deviation (SD_{rel}) for one, a series, or all of the replicate analyses you performed: $SD_{rel} = SD/average\ value \times 100$. It is expressed as a percentage: $SD_{rel} \leq 10\%$ is considered acceptable and $SD \leq 5\%$ is excellent.

Name _____ Date _____

Report

1. Plot the calibration chart for the alcohol determination from the data in step 3.

2. List all the alcohol specimens chromatographed with their retention times.

3. Record your data in Table 35.1.

TABLE 35.1
Record of Data

Percent Ethyl Alcohol in Standard	Area of Ethyl Alcohol Peak	Area of Internal Standard Peak	Ratio of Ethyl Alcohol Area/Internal Standard Area

4. What is your unknown sample identification? What is the percent ethyl alcohol in your unknown?

5. If you were instructed to use peak heights as an option: Draw a table similar to Table 35.1 in your notebook. Replace the area terms with peak height data. Calculate the ratio. Prepare a calibration curve. Redetermine the ethyl alcohol amount using peak heights and report below.

6. If your teacher supplies the true value for your unknown, calculate the percent error for both the peak area and peak height methods. Which method seems better and why?

7. Report the SDrel for the data your instructor required you to calculate.

8. If instructed to do so, make photocopies of your chromatograms and include them with your report.

Experiment

36

Forensic Entomology

Teaching Goals

The goals of the lesson are to allow the student to become familiar with some of the basic techniques and methodologies used by forensic entomologists to collect specimens, identify blow flies, and estimate time of death. Each student will gain additional hands-on experience with the collection, preservation, and study of insect specimens. After this lesson, each student will know how to identify various adult blow flies, their eggs, pupae, and larval stages and how to utilize the data acquired from the students' study of the life cycles of common blow flies to estimate time of death.

Background Knowledge

Blow flies are a large, varied group of dipterous insects of the family Calliphoridae that deposit their eggs on carrion and human remains. Various species of blow flies are found throughout the world. The majority of adult blow flies' bodies range from 6 to 14 mm in length and have a blue, green, bronze, or black metallic luster. Blow flies are one of the most important species that provide data relating to the accurate estimation of postmortem time intervals; see Figures 36.1 and 36.2.[1]

The biology of the blow fly varies among its many species. Environmental conditions such as the weather, temperature, humidity, amount of shade or sunshine, and conditions of indoor or outdoor habitats help to dictate their development. The black blow fly is a cool weather insect; its adult stage thrives in the winter and is most abundant in the onset of spring or in the cool days of autumn. Yet, other species of blow flies are abundant during the months of summer, when conditions are warm and humid. Still other species prefer to live indoors rather than outdoors.

Since the developmental stages of blow flies are highly known and predictable at ambient temperatures, blow flies are considered a valuable tool in forensic science. Mature female blow flies usually arrive within minutes of human death because they have the ability to smell the death of a body from many miles away. Each female blow fly lays eggs in the normal orifices of the body and in exposed, open wounds. Generally, a blow fly deposits thousands of eggs on carrion over their short life span, which is typically 2 to 8 weeks long. The egg deposits normally contain from 50 to 1,000 eggs, and incubation occurs in less than a day in warm and humid conditions.

On average, the eggs hatch into first-stage maggots within 24 hours. These feed and then molt into second-stage maggots, which feed for several hours and then molt into third-stage maggots. Masses of third-stage maggots may produce heat, which can raise the temperature around them more than 10°C. After additional feeding, the third-stage maggots move away from the body and typically burrow in the upper centimeters of the soil and pupate for up to a week, at which time they undergo metamorphosis into adult flies.

Adult flies emerge from pupation and make their way to the soil surface. Depending on temperature and the substrate upon which they are feeding, maggots usually complete development within 4 to 10 days.

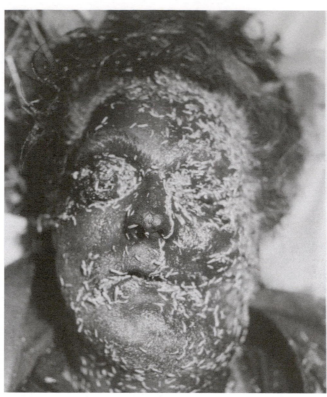

Figure 36.1
Homicide investigation in which the victim is covered with blow flies. From Reference 1.

1.0mm

Figure 36.2
An adult blow fly. Copyright U.S. National Library of Medicine, 8600 Rockville Pike, Bethesda, MD 20894. A full-color version of this figure can be obtained from http://www.nlm.nih.gov/visibleproofs/.

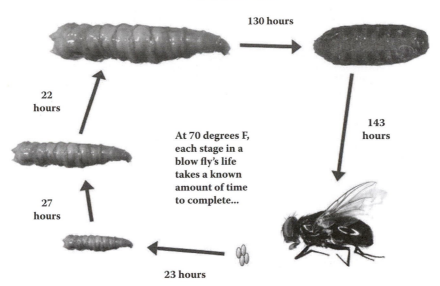

The blow fly life cycle has six parts: the egg, three larval stages, the pupa, and adult.

130 hours

22 hours

143 hours

At 70 degrees F, each stage in a blow fly's life takes a known amount of time to complete...

27 hours

23 hours

Figure 36.3
Life cycle of a blow fly. Copyright U.S. National Library of Medicine, 8600 Rockville Pike, Bethesda, MD 20894. A full-color version of this figure can be obtained from http://www.nlm.nih.gov/visibleproofs/.

About 1 week later, females begin to deposit eggs, and the life cycle is repeated. Blow flies usually develop from egg to adult in only 10 to 25 days and complete four to eight generations each year. The complete life cycle of the blow fly can be seen in Figure 36.3.[1,2]

Equipment and Supplies

1. Three (3) disposable 8-in. plastic dishes with a sealable top
2. One pound of ground beef
3. One small bag of potting soil (sterile preferred)
4. One gingerbread man cookie cutter
5. A 32-oz. bottle of ethanol (methanol or isopropyl if ethanol is not available)
6. Soft, featherweight forceps (to handle insects)
7. Fine-tip tweezers
8. A dozen small 2- to 4-oz. heavy paper cups, glass or plastic containers with tops
9. Stereomicroscope or magnifying glass
10. Black fine-point Sharpie® marker
11. A 6-in. metric ruler
12. Small trowel
13. One 8-oz. dry measuring cup
14. No. 2 specimen pins and specimen mounts
15. A set of Ward's Natural Science® Forensic Insect Identification Cards

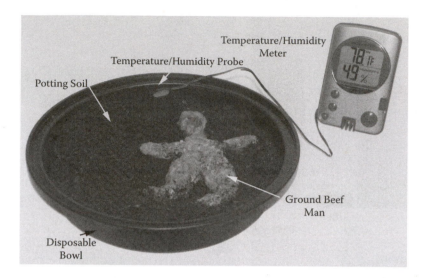

Figure 36.4

A completed specimen container ready for placement into desired environment. The portable meter is used to collect temperature and humidity data over the course of observation. The raw beef cutout can be exchanged with a chicken leg.

Procedure

Students working in groups of three or four will acquire all the required materials from the instructor. One student will be chosen as team leader. The remaining students, following the instructions of the team leader, will prepare three specimen collection apparatuses as follows:

1. Place three of the 8-in. containers on a lab table.

2. Each student will place six cups of dry potting soil into one of the 8-in. diameter containers.

3. One student will mold the ground beef into a flat, 6-in. wide × 4-in. high × 3/8-in. thick rectangular-shaped patty.

4. Each student will cut out one ground beef specimen with the cookie cutter.

5. Each student will place his or her specimen on top of the potting soil in their container; see Figure 36.4.

6. Each team of students will place their specimens in three different locations:
   • One outdoors in direct sunlight
   • One outdoors in a shady area under a bush
   • One inside in a shed, garage, or storage area

7. Each student will monitor his or her specimen container daily over a 1- to 2-week period and record his or her observations on a data collection sheet; see Appendix A.

8. At the end of the observation period, any insect life is preserved by pouring a cup or two of alcohol directly on top of the specimen and surrounding area.

9. Each container is sealed with a top and transported to the laboratory for observation.

10. Each student will sort out the various stages of blow fly (eggs, adult flies, larval stages, and pupae) with feather-weight forceps and place each category into separate heavy paper cups containing 1 oz. of alcohol.

11. After the sorting process is complete, the individual blow flies, maggots, and pupae are compared to those in the Forensic Insect Identification Cards reference set (a stereomicroscope or magnifying glass should be used to make observations).

12. Selected specimens should be mounted for study.

13. A tentative identification of each species of blow fly is rendered.

14. The data are examined, studied, and tabulated, and all of the questions in the report section should be answered based on your observations and collected data; see Appendix A, Appendix B, and Ward's Forensic Insect Identification Card Set.

Name ————————————————— Date ——————————————————

Report

1. Describe the appearance of the following:

   a. Blow fly eggs

   b. Blow fly larvae

   c. Blow fly pupae

   d. Blow fly adults

2. Does the temperature, humidity, or weather affect the types of blow flies that appear?

3. Does the environment affect the types of blow flies that appear?

4. Using your empirical observation and the data your group collected over the course of this exercise, which species of blow fly prefers the following conditions:

a. Outdoors in the shade

b. Outdoors in full sunlight

c. Indoors

d. Hot and humid

e. Cool and dry

APPENDIX A

Forensic Entomology Data Sheet - Blow Flies

*Life Cycle Blow Fly

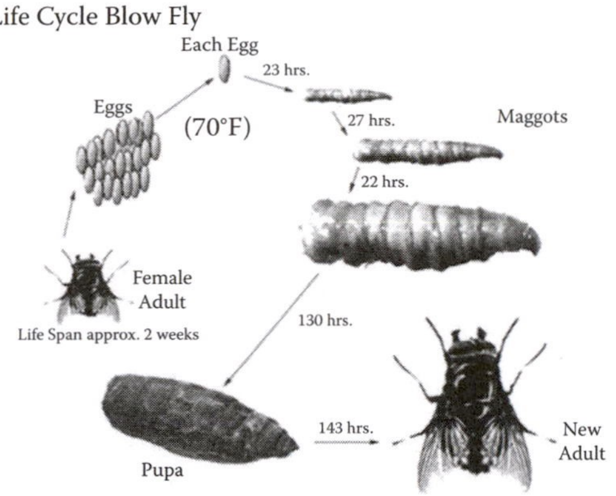

Sketch		
I.D.	Order	_____
	Family	_____
	Species	_____

Circle/Check/Fill-in

Case No. _____ Date _____ Time _____

Location _____ Substrate _____

Environment   Inside   Outside   Sunshine   Shady   Cloudy   Wet   Dry

Weather Conditions _____

Temperature _____ °F & _____ °C   % Humidity _____

Surface Condition _____

Blow Fly Activity   Yes   No

		Metallic Body Color	Numbers	
☐ Eggs	Stage		☐	Adults
☐ Maggots	1 2 3	Blue Green Black	☐	Eggs
☐ Pupae		Other _____	☐	Maggots
			☐	Pupae

Physical Data: Length _____ in./mm   Length Range _____

Body Shape _____ Texture _____ Body Color _____

Color Variation _____

Wing Size _____ Wing Appearance _____

Head Marking _____ Body Markings _____

Hair on Body   Absent   Light   Moderate   Heavy

Eye Color _____ Eye Size _____ Eye Shape _____

*Life cycle of blow fly based on: http://www.nlm.nih.gov/visibleproofs/visit/, Copyright, Privacy, Accessibility U.S National Library of Medicine, National Institutes of Health, Health & Human Services.

Figure A36.1

APPENDIX B

Forensic Entomology

- COLLECTION
- PRESERVATION
- DOCUMENTATION
- KILL SOME SAMPLES AND RETAIN SOME FOR GROWTH
- GROWTH WITH MEDIA (FOOD)
- GROWTH PATTERNS, SCIENTIFICALLY DATING OF SAMPLES RECOVERED

Figure B36.1

Blue Blow Fly
Calliphora vomitoria

Blue Blow Flies like sunny exterior shady locations

Figure B36.2

Green Blow Fly *Phanicia sericata*

Green Blow Flies like sunny exterior locations

Figure B36.3

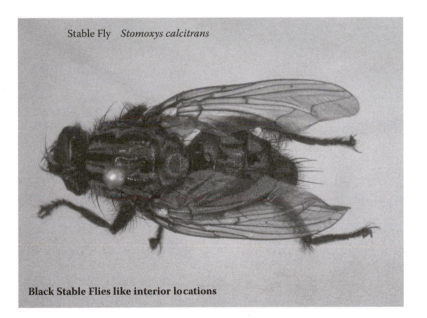

Stable Fly *Stomoxys calcitrans*

Black Stable Flies like interior locations

Figure B36.4

References

1. Byrd, J.H., and Castner, J.L., *Forensic Entomology, The Utility of Arthropods in Legal Investigations*, CRC Press, Boca Raton, FL, 2001.
2. U.S. National Library of Medicine, National Institutes of Health, U.S. Department of Health and Human Services, Visible Proofs: Forensic Views of the Body, http://www.nlm.nih.gov/visibleproofs/, U.S. National Library of Medicine, Bethesda, MD.

Experiment **37**

Basic Ballistics

Teaching Goals

The goals of the lesson are to allow the student to become familiar with some of the basic techniques and methodologies used by forensic ballistics specialists to examine firing pin and breech mark impressions on cartridge casings. After this lesson, each student will know how to identify and compare firing pin and breech mark impressions.

Background Knowledge

Impression marks are caused by the action of the mechanisms of a firearm during the firing process. When an automatic pistol is discharged, the shooter pulls the trigger until its hammer mechanism strikes the firing pin. In turn, the firing pin strikes the cartridge's primer, initiating the burning of the smokeless powder charge. The smokeless powder burns, producing gas and pressure, which pushes the bullet out of the casing and through the pistol's barrel. After passing through the length of the barrel, the bullet exits the pistol's muzzle accompanied by the remnants of the burning smokeless powder, primer residues, and flame or muzzle flash. Simultaneously, the pistol's slide is forced backward, which causes the spent casing to exit the pistol's ejection port. Figure 37.1 shows the basic components of an automatic pistol. Figure 37.2 depicts a discharging pistol.

The movement of the pistol's mechanisms, the shell or casing, and the firing pin cause several tool marks to be imparted onto the bullet and casing. The bullet acquires land and groove marks as well as minute striations as it passes through the barrel and interacts with its rifling. The shell receives a firing pin impression when the pistol's firing pin hits the primer; breech marks when the explosion forces the head of the casing backward into the pistol's breech; and extractor marks when the pistol's extractor mechanism grips the shell's rim and head. Figure 37.3 shows the inside of a pistol.

Firing pin impressions are produced when the firing pin strikes the primer of a center-fire cartridge case or the rim of a rim-fire cartridge case. If the nose of the firing pin has surface marks or damage, a negative impression of these imperfections gets stamped into the softer metal of the primer or rim shell casing. Figure 37.4 shows the nomenclature of a cartridge along with before and after illustrations of a primer cap. Figure 37.5 depicts a firing pin mechanism found on revolvers. The comparison image in Figure 37.6a shows a firing pin impression on one center-fire cartridge case on the right and a firing pin impression on a rim-fire cartridge case on the left. The images in Figure 37.6b show ejector marks on a discharged automatic shell casing.

Ejector marks are typically produced when a pistol's action ejects the cartridge from its port during the firing process. Ejector marks like the one shown can only be reproduced when a pistol is discharged and not simply by hand chambering and ejecting a live cartridge.

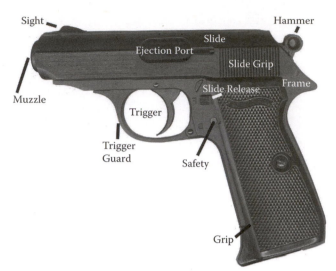

Figure 37.1
A typical automatic pistol showing some of its components.

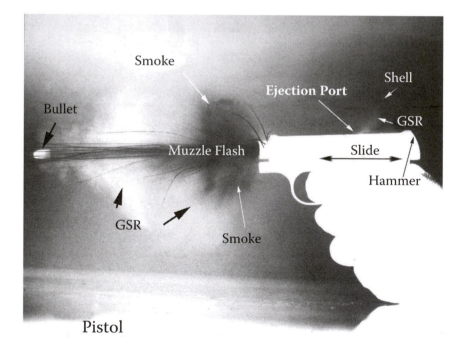

Figure 37.2
A discharging pistol showing the bullet, muzzle flash, gunshot residues (GSR), and ejected shell.

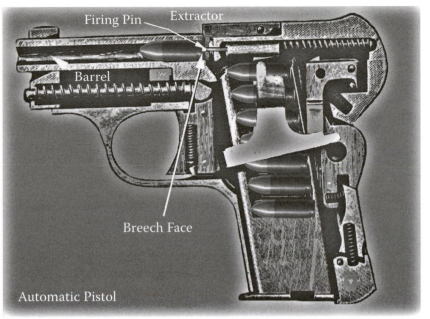

Firing Pin   Extractor

Barrel

Breech Face

Automatic Pistol

Figure 37.3
A cutout view of an automatic pistol.

Breech Marks

Breech marks (Figure 37.7) are another impression mark commonly caused by the firearm's action on the cartridge case. When a pistol is discharged very high pressures are generated.

Equipment and Supplies

1. A 6-in. metric rule
2. Protractor
3. Magnifying glass
4. Pencils
5. One set of fine-point Sharpie® colored marking pens
6. Black fine-point Sharpie marker
7. A set of Ward's Natural Science® Ballistics Cards

Procedure

- Each student should read the background knowledge section of the laboratory manual. Next, review the set of Ward's Ballistic Cards to see color examples of firing pin and breech mark impressions.
- Next, each student working alone will acquire all the required materials from the instructor.
- Each student should carefully examine the 12 casing heads (Figure 37.8). Make linear and angular measurements when necessary. Note the data directly on Figure 37.8. Circle the matching class characteristics and accidental characteristic patterns found on each casing head with the same color marker.

Cartridge Nomenclature

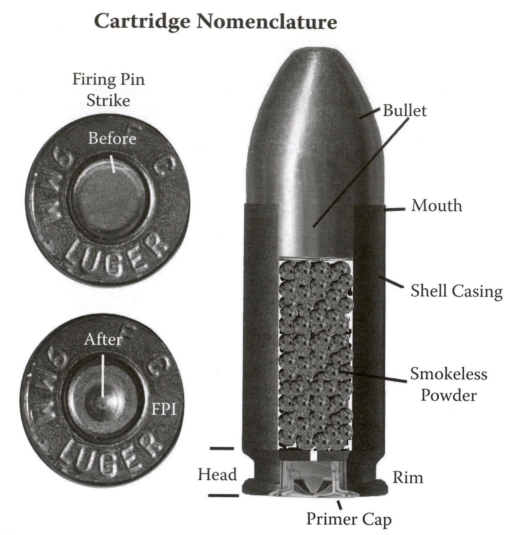

Figure 37.4
A cutout view of a cartridge case showing its components as well as before and after views of its center-fire cartridge case. FPI, firing pin impression.

- Using your collected data, determine the matching and nonmatching groups of cartridge casings. Circle the casing heads with matching firing pin and breech mark impressions.
- Note the matching and nonmatching specimen cartridge casings in Table 37.1.
- Answer all questions in the Report section
- Check your answers with your instructor.
- If your answers are correct, proceed to Experiment 38.

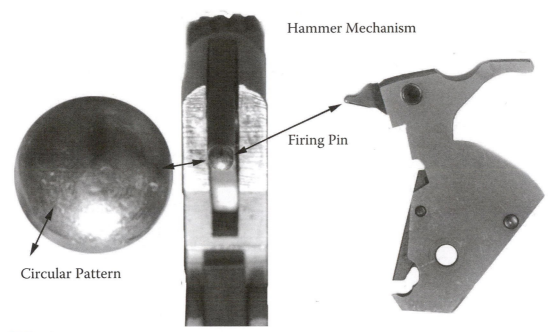

Figure 37.5
A typical firing pin used for revolvers. The firing pin is incorporated into the hammer mechanism. Note the circular patterns of striations on the tip of the firing pin.

Figure 37.6a
A rim-fire cartridge case with a firing pin impression (left) and a center-fire cartridge case with a firing pin impression (right).

Figure 37.6b
An image of ejector marks on the left and an enlargement of the same impression on the right. Note the parallel striations in the enlarged image.

Cartridge Case Head

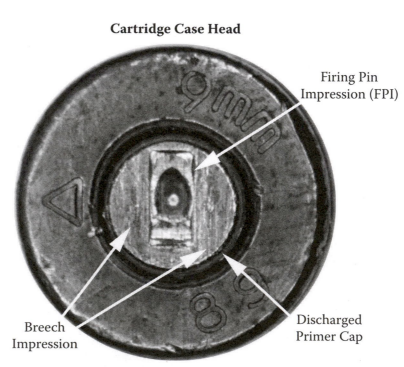

Firing Pin
Impression (FPI)

Breech
Impression

Discharged
Primer Cap

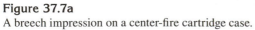

Figure 37.7a
A breech impression on a center-fire cartridge case.

Figure 37.7b
A close-up photograph of the breech mark depicted in Figure 37.7a.

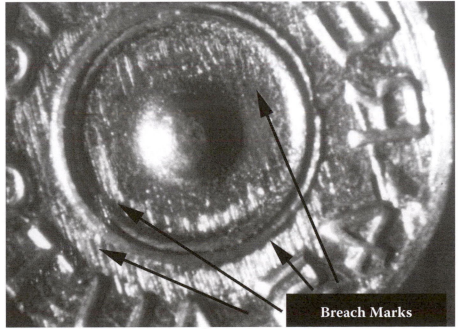

Figure 37.7c
An example of a revolver breech mark.

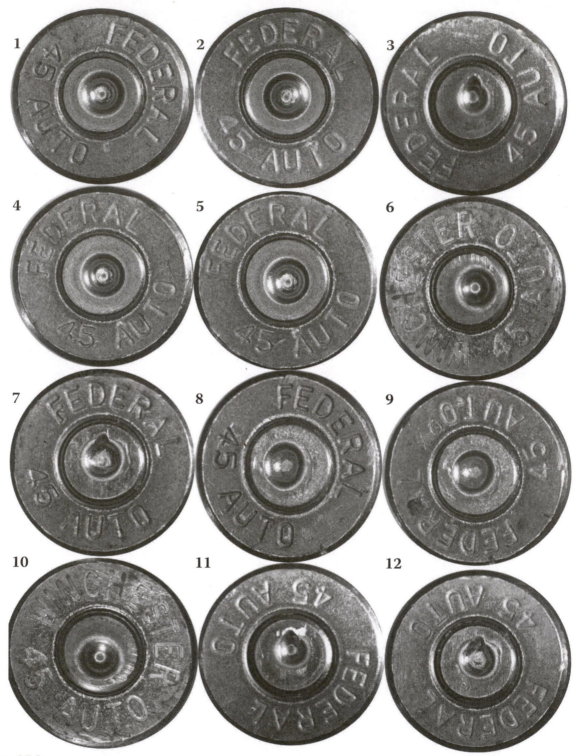

Figure 37.8
Review the 12 images and choose the matching sets of firing pins. Carefully examine the 12 casing heads. Make linear and angular measurements when necessary. Circle the matching class characteristic patterns and accidental characteristics of each casing head with the same color marker. Note the data in the tabulation sheet. Using your collected data to determine the matching and nonmatching groups of cartridge casings, circle the casing heads with the matching breech marks and firing pin impressions. Note the number of the matching and nonmatching cartridge casing heads in Table 37.1.

TABLE 37.1
Tabulation Table: Note the Number of the Matching and Nonmatching Cartridge Casing Heads

Specimen no.	Matching Groups	Nonmatching Groups
1		
2		
3		
4		
5		
6		
7		
8		
9		
10		
11		
12		

Name _____ Date _____

Report

1. How many casing heads matched specimen No. 1 in Figure 37.8? Make a sketch of each of the matching specimens.

2. Make a sketch of the firing pin impression in Figure 37.7b.

3. Do any of the firing pin impressions in Figure 37.8 match the firing pin impression depicted in Figure 37.7? If yes, make a sketch of each matching impression.

4. Using your empirical observation and the data you collected over the course of this exercise, how many firearms were used to discharge all of the cartridges depicted in Figure 37.8?

References

1. Hatcher, J.S., Jury, F.J., and Weller, J., *Firearms Investigation, Identification and Evidence*, 2nd ed., Stockpole Co., Harrisburg, PA, January 1977.
2. Svensson, O., and Wendel, A., *Techniques of Crime Scene Investigation*, 5th ed., Elsevier, New York, 1971.
3. Rowe, W.F., Firearm and Tool Mark Examinations, in *Forensic Science*, edited by S.H. James and J.J. Nordby, CRC Press, Boca Raton, FL, 2003, pp. 327–343.

Experiment **38**

Basic Ballistics 2

Teaching Goals

The goals of the lesson are to allow the student to become familiar with some of the basic techniques and methodologies used by forensic ballistics specialists to examine firing pin and breech mark impressions on cartridge casings. After this lesson, each student will know how to identify and match firing pin and breech mark impressions.

Background Knowledge

A body is discovered in a wooded area of a local park by two persons walking their dogs at 6:00 a.m. on a Saturday morning. The body is found under bushes near some drag marks.

One of the dog walkers calls 911 to report their findings. The 911 operator sends a police patrol car to the location to investigate. The first officer on the scene examines the body and establishes that the person is dead; the officer immediately secures the crime scene and calls for the duty detective and the medical examiner. The duty detective, on arriving at the scene, asks for the crime scene unit (CSU) to respond.

A search of the crime scene reveals the presence of bloodstain patterns, two questionable footwear prints, evidence that the victim was shot numerous times in the head and body, 12 discharged (spent) cartridge casings, and one 0.45-caliber automatic Smith and Wesson pistol several feet away from the body. The pistol still had four live 0.45-caliber rounds: three in its magazine and one in its chamber. The collected firearm and spent cartridge casings were sent to the ballistics unit for examination. A picture of each of the 12 spent casings found by CSU personnel, in their order of acquisition, is shown in images CSU-1 to CSU-12.

During the investigation, information on several suspects developed. The investigating officers, in cooperation with the district attorney's office, presented their affidavits for search warrants to a state Supreme Court justice. On determining that there was probable cause to search the suspects' residences, state Supreme Court Justice John Smith issued a search warrant for each suspect's home. During the searches, one firearm was confiscated from each of two of the locations (one 0.45-caliber Smith and Wesson™ pistol, one 0.40-caliber Colt™ pistol), and two firearms were obtained from the third residence (one 0.40-caliber Winchester pistol and one 9-mm Ruger™ pistol). All four firearms were sent to the ballistics examiner for comparison to the cartridge cases collected at the crime scene.

The ballistics examiner prepared five sets of known test specimens of cartridge cases with the five firearms collected at the crime scene and during the searches: one 0.45-caliber Smith and Wesson pistol recovered at the crime scene and the four firearms obtained from the suspects' homes. Next, each student shall examine the photographs of each set of known specimens and than compare each to the photographs of the cartridge cases collected at the crime scene (CSU-1 through CSU-12). Finally, each student shall report the results of his or her examinations and comparisons in the report portion of this laboratory exercise.

Figure 38.1
CSU-1

Figure 38.2
CSU-2

Figure 38.3
CSU-3

Figure 38.4
CSU-4

Figure 38.5
CSU-5

Figure 38.6
CSU-6

Figure 38.7
CSU-7

Figure 38.8
CSU-8

Figure 38.9
CSU-9

Figure 38.10
CSU-10

Figure 38.11
CSU-11

Figure 38.12
CSU-12

Figure 38.13
Two cartridge casings fired with known gun No. 1, a 0.45-caliber Smith and Wesson pistol.

Figure 38.14
Four cartridge casings fired with known gun No. 2, a 0.45-caliber Smith and Wesson pistol.

Figure 38.15
Three cartridge casings fired with known gun No. 3, a 0.40-caliber Colt pistol.

Figure 38.16
Four cartridge casings fired with known gun No. 4, a 0.40-caliber Winchester pistol.

Figure 38.17
Three cartridge casings fired with known gun No. 5, a 9-mm Ruger pistol.

Name _____ Date _____

Report

1. After studying the firing pin, ejector mark, and breech mark impressions on the head of each cartridge casing, how many firearms do you think were discharged at the crime scene?

2. How many cartridge casings recovered at the crime scene were discharged by the firearm used to prepare the known set for Gun No. 1 (Figure 38.13)? Sketch the matching patterns on the questioned and known cartridge casings.

3. How many cartridge casings recovered at the crime scene were discharged by the firearm used to prepare the known set for Gun No. 2 (Figure 38.14)? Sketch the matching patterns on the questioned and known cartridge casings.

4. How many cartridge casings recovered at the crime scene were discharged by the firearm used to prepare the known set for Gun No. 3 (Figure 38.15)? Sketch the matching patterns on the questioned and known cartridge casings.

5. How many cartridge casings recovered at the crime scene were discharged by the firearm used to prepare the known set for Gun No. 4 (Figure 38.16)? Sketch the matching patterns on the questioned and known cartridge casings.

6. How many cartridge casings recovered at the crime scene were discharged by the firearm used to prepare the known set for Gun No. 5 (Figure 38.17)? Sketch the matching patterns on the questioned and known cartridge casings.

7. How many of the cartridge casings collected at the crime scene were not associated to one of the firearms collected in this investigation?

Crime Scene Drawing with Microsoft Word®

Teaching Goals

After this lesson, each student will be able to use the Microsoft Word® drawing toolbar to prepare a professional-looking, accurate, to-scale drawing of a crime scene.

Background Knowledge

The need for preparing professional-looking finished sketches from the notes, measurements, and rough sketches made at the crime scene is an essential aspect of a crime scene investigator's work. These finished diagrams and drawings are used to help reconstruct the event. They will also be presented in court during testimony and throughout the trial to aid the court and jury to better understand the evidence and events of the case. Thus, the diagrams must be clear, accurate, and easy to read and understand. Computer-aided design (CAD) programs are available for this purpose. However, these types of specialty programs are typically very costly and require a great deal of formal training to use. Reference 1 is a recent journal article that explains a simple way to make professional-looking crime scene diagrams and drawings using a program accessible to most people: Microsoft Word.

Materials

1. A prepared mock crime scene.
2. A rough sketch of a crime scene. Appendix A has a few examples of rough sketches and examples of measurement methods (see Figures 39.9 to 39.14).
3. Notes and measurements from a previous crime scene, your rough sketch, and notes taken during your mock crime scene laboratory Experiment 8 or your assigned rough sketch and notes given to you by your instructor.
4. A current version of Microsoft Word and a computer.
5. A 30°/60°/90° triangle ruler with both metric and standard dimensions.
6. Using the procedure described draw the desired crime scene using the standard measurement system used in the United States. Then, prepare the same diagram using the metric system. Which was easier to prepare?

Procedure

To prepare a finished, to-scale crime scene drawing from your rough sketch and notes taken during your mock crime scene laboratory Experiment 8 or your assigned rough sketch and notes, follow the instructions given here. For a detailed explanation of how to use the drawing toolbar in Microsoft Word to accomplish this exercise, refer to the article by Lamarche.[1]

Toolbar Option Selections

1. Open Microsoft Word.
2. Open Tools and click on Options; a File dialogue box will appear.
3. Open **General tab** and make sure the following boxes are checked and settings made:
    a. Provide feedback with animation
    b. Update automatic links at open
    c. Mail as attachment
    d. Recently used file list set to *4 entries*
    e. Measurement units set to *inches*
    f. Make certain that the "Automatically create drawing canvas…" **is not checked**.
4. On the File dialogue box, click the **Print tab** and check the *Drawing Objects* box.
5. Finally, on the File dialogue box, click the **View tab**; under *Print and Web Layout Option,* check the *Drawings* box.

File: Page Setup Options

1. Open File on toolbar and choose **Paper tab**; set to the *maximum* paper size your printer will allow, typically 8½ × 14 in.
2. Next, choose **Margin tab** and set to Landscape mode.

View: Toolbar Setup Options

1. Open View on toolbar and choose **Toolbar tab** and check *Standard, Formatting, Drawing,* and *PDF* boxes.
2. Next, under View, check the *Ruler* box.

Drawing Toolbar Settings and Options

1. The **Drawing toolbar** appears on the top of the bottom border of the page and contains the following numerous iconic buttons, from left to right:
    • Draw button: controls grid sizing and manipulation
    • White arrow button: used to select objects
    • Rotation button: used to rotate grid text or objects
    • AutoShapes button: used to select lines, objects, and design elements
    • Line button: used to insert lines
    • Black arrow: used to draw arrows
    • Rectangular or oval button: used to draw rectangles or ovals, respectively
    • Text button and Word art button: used to create or insert text
    • Diagram button: used to insert diagrams from files
    • Clip art button: used to insert clip art or pictures from files

- Paint bucket button: used to fill objects or space with color
- Brush button: used to paint lines and objects
- "A" button: used to select desired color
- Solid lines, broken lines, and arrow buttons: used to change styles
- Square button: used for applying shadows
- Cube button: used to make 3D shapes

2. Some button styles and functions may vary with different versions of Microsoft Word.

Steps in Preparing a Crime Scene Drawing

1. Acquire the rough crime scene sketch and notes to be used in this exercise. Study the rough sketch to determine the overall dimensions of the final drawing.

2. Set up your printer to print legal size paper (8½ × 14 in.) in landscape mode, orientation to landscape, and the printable area to standard/center; see Figure 39.1.

3. Set the page setup margins to default settings as demonstrated in Figure 39.2.

4. Set the page setup layout screen menu to header and footer 0.5 in. and the vertical alignment to *Top* as demonstrated in Figure 39.3.

5. Click the draw tab on the drawing toolbar and activate the drawing grid screen menu. Next, check the two snap to buttons; check the use margins box and make the horizontal spacing and the vertical spacing 0.15 in. Finally, check the display gridlines on screen box and vertical every box (see Figure 39.4).

6. Establish a drawing ratio and scale. Set the page setup layout tab border button to box (see Figure 39.5 for resulting canvas). Each small grid box on the finished canvas is equal to 3 in. The scale of 1 in. on the canvas is equal to 1 ft. Thus, four small grid boxes are equal to 1 ft in any horizontal and vertical direction on the finished canvas. When transferring measurements the 30°/60°/90° triangle ruler can be placed directly onto the screen to aid in making accurate, to-scale measurements and measurement transfers. The grid can be removed or added to the finished diagram at any time by simply clicking the grid display guideline on screen box in the drawing grid popup menu (see Figure 39.4).

7. To add pictures, geometric shapes, lines, flowcharts, clip art, or prepared shapes or objects to the finished drawing, activate the AutoShapes tab on the draw toolbar and click the desired option on the screen menu. Next, place the mouse arrow over the desired object, shape, and so on and insert it onto the canvas. This is accomplished by clicking the left mouse button and, while holding it down, dragging the object to the desired location on the canvas. Each option on the AutoShapes menu allows for formatting of geometric shapes, lines, pictures, clip art, and so on. The Format AutoShape screen in Figure 39.6 illustrates one of the many available option menus for changing the size, shape, thickness, color, transparency, and so on of a line, object, clip art, or the like. All of these options and more are available for adjusting the appearance, size, color, or shape of any text, item, or object. Figure 39.7 demonstrates the use of more AutoShape options and the insertion of ready-drawn clip art onto the canvas. Figure 39.8 depicts a completed crime scene drawing prepared with Microsoft Word drawing toolbar.

With patience and practice anyone can prepare professional-looking, full-color, accurate, to-scale crime scene drawings using this method. The interested reader is referred to the complete article by Lamarche[1] and any help text when using Microsoft Word for a detailed description of this method.

Reference

1. S. Lamarche, Microsoft Word Crime Scene Drawing, *J. Forensic Ident.*, 57, 848–869, 2007.

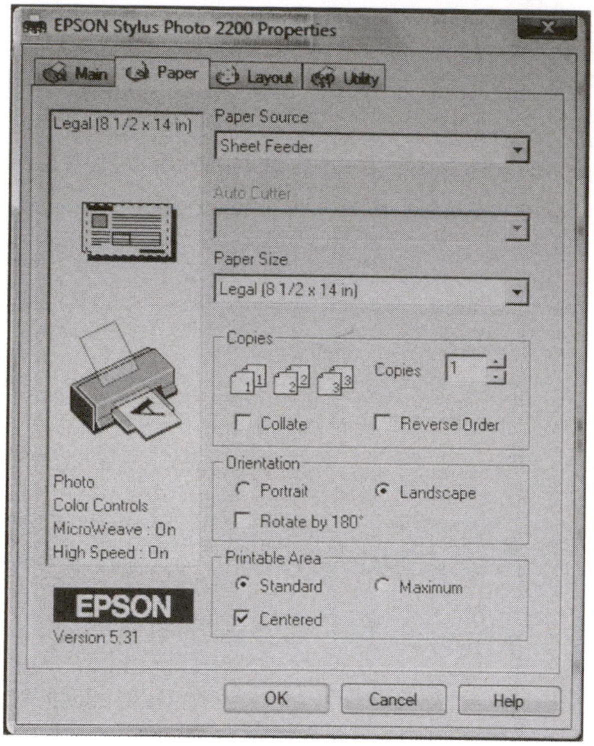

Figure 39.1
The paper setup window for available printer is shown above. Set the paper to legal size, the orientation to landscape, and the printable area to standard/center.

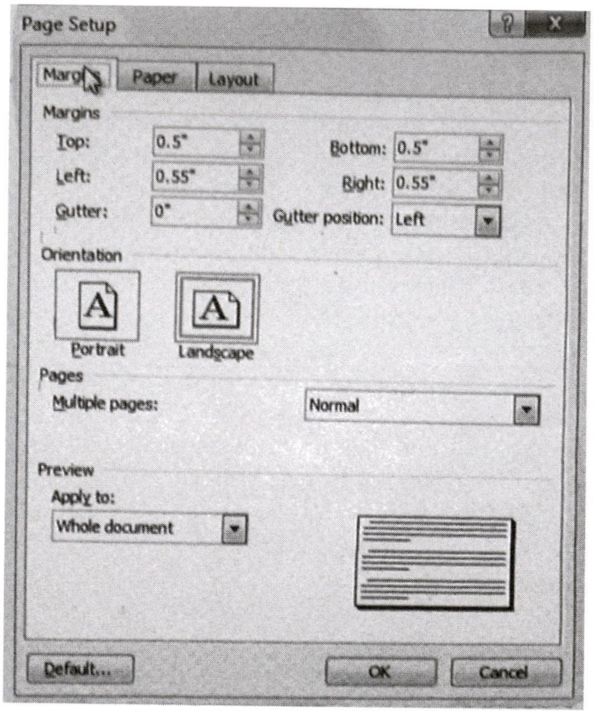

Figure 39.2
On the page setup screen, set the margins to default values as illustrated.

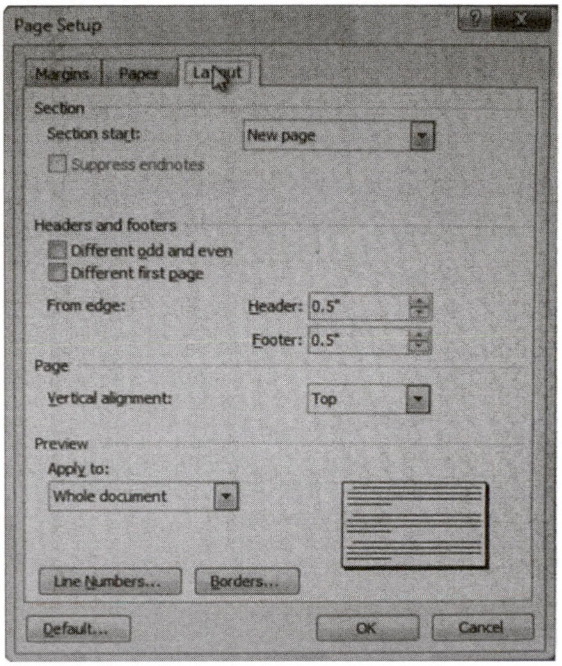

Figure 39.3
On the page setup screen menu, set the layout tab to header 0.5 in. and to footer 0.5 in. and the vertical alignment to *Top*.

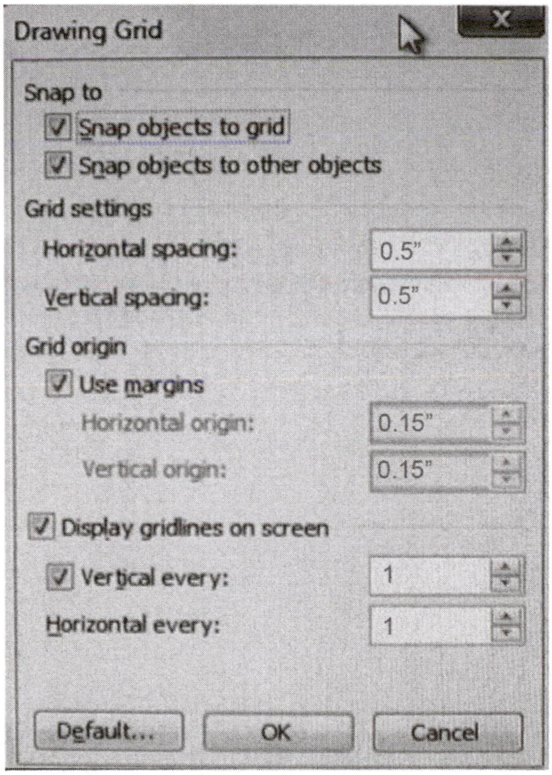

Figure 39.4
Click the draw tab on the drawing toolbar and activate the drawing grid screen menu. Next, check the two snap to buttons; check the use margins box and make the horizontal spacing and the vertical spacing 0.15 in. Finally, check the display gridlines on screen box and vertical every box.

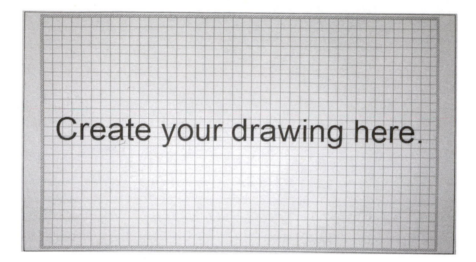

Figure 39.5
The resulting canvas with each small grid box equal to a 3 × 3 in. square, and the scale is 1 in. on the canvas equals 1 ft.

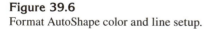

Figure 39.6
Format AutoShape color and line setup.

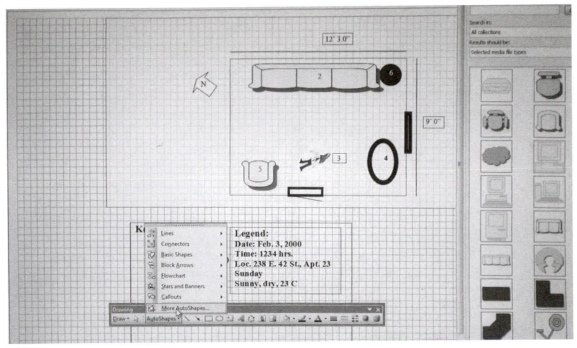

Figure 39.7
Insert prepared shapes, photographs, text, and so on using draw toolbar and appropriate box, tab, or icon.

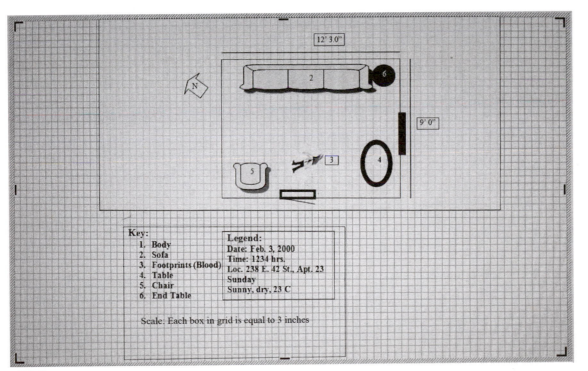

Figure 39.8
A finished drawing prepared with Microsoft Word documents toolbar, rough sketch, and notes from Experiment 8 crime scene.

APPENDIX A

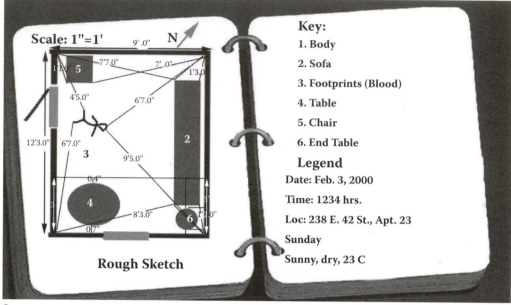

Figure 39.9
A rough sketch with notes and measurements from Experiment 8 crime scene investigation.

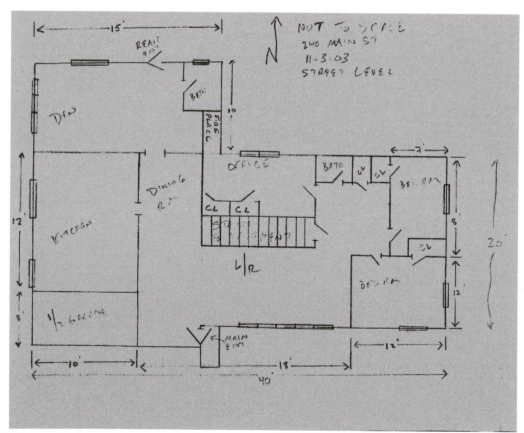

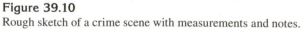

Figure 39.10
Rough sketch of a crime scene with measurements and notes.

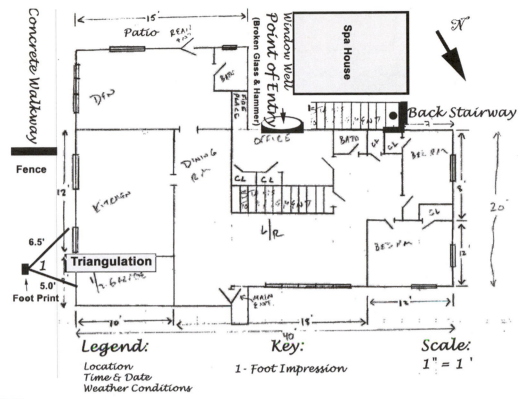

Figure 39.11
A more detailed example of a rough sketch of a crime scene with measurements and notes.

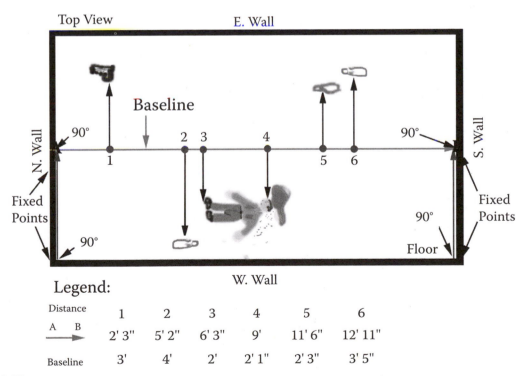

Figure 39.12
A rough sketch of crime scene with baseline measurements.

Legend:

Distance A → B	1	2	3	4	5	6
	2' 3"	5' 2"	6' 3"	9'	11' 6"	12' 11"
Baseline	3'	4'	2'	2' 1"	2' 3"	3' 5"

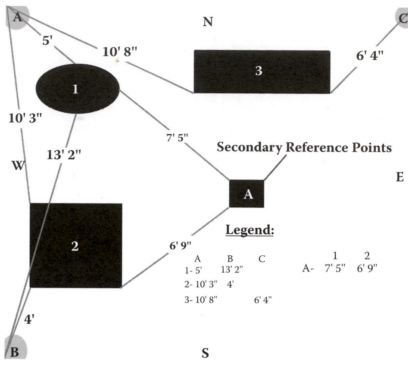

Figure 39.13
A rough sketch of crime scene with triangulation measurements.

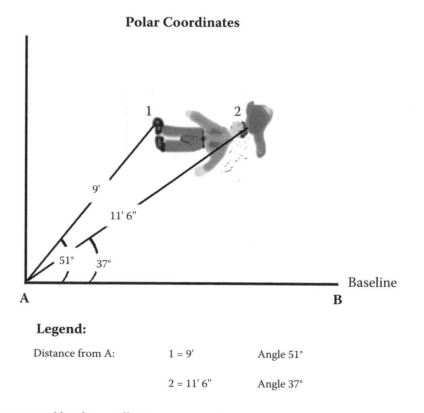

Figure 39.14
A rough sketch of crime scene with polar coordinate measurements.

Appendix

Reference Data for Polymer Films

This section contains two series of infrared spectra of polymers. The first, labeled Spectra 1 through 29, are the spectra of the specific polymers from a spectral database. The identity of each particular polymer can be found in the upper left-hand corner of each chart. These can be considered quality spectra. The very observant student may notice that Spectra 20 and 21 are identified as the same polymer, poly(methyl methacrylate) but have slightly different spectral curves. This is because they were produced by separate companies and polymerized slightly differently but remain the same generic class of polymer. The second are a series of spectra collected from various packaging polymer films and determined as thin films employing the method described in Experiment 32. These have been identified as to the source of the film and the best computer match of the film to a reference database for polymer identification. This information can be found in the upper left-hand corner of the chart. These are labeled Polymer Wrap 1 through 12. A listing of spectra and polymer wraps may be found at the end of this section.

When reviewing the spectra of a polymer film mounted in air as in Experiment 32, the analyst may notice what appears to be a sinusoidal wave superimposed on the spectra. Please compare Spectra 13 for linear polyethylene polymer to Polymer Wrap 11. Notice the three arrows that point to this superimposed wave. These arrows merely direct one's attention to part of the total wave. This wave is an interference pattern that results from the passing of the infrared beam across interfaces of very different refractive indices (air-polymer film-air). Each of the polymer films have this interference wave superimposed upon the spectrum of the film in lesser or greater degrees. In the remainder of the Polymer Wrap spectra only one arrow is employed to delineate the pattern where it is easily noticeable. It should be recognized that this pattern did not cause any major difficulty for the computer search software in its identification of the most probable chemical composition of the wrapping film.

Students are welcome to employ either set of spectra to assist them in the identification of polymer films that the instructor supplied for identification.

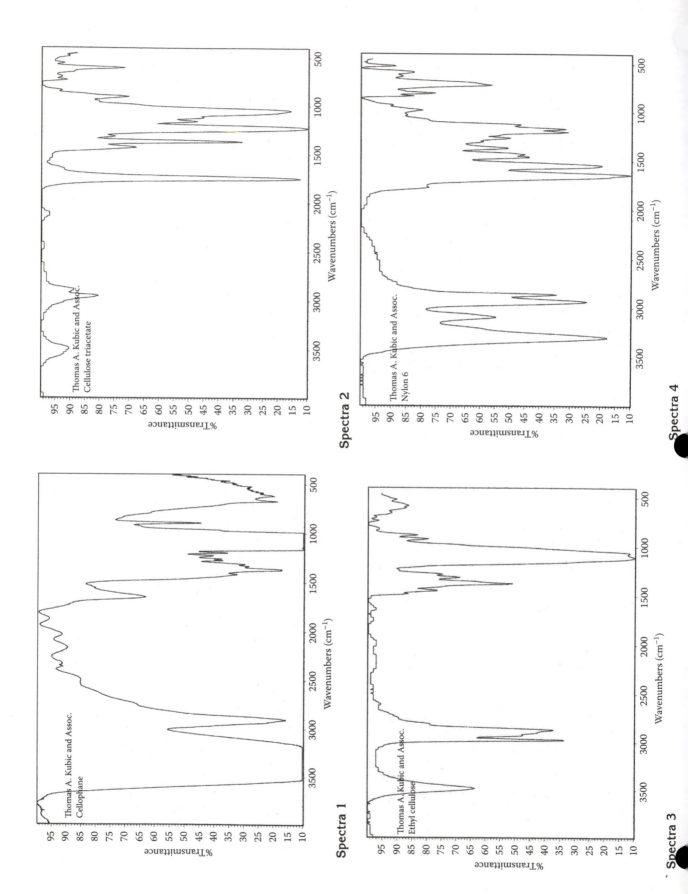

Spectra 1

Thomas A. Kubic and Assoc.
Cellophane

Spectra 2

Thomas A. Kubic and Assoc.
Nylon 6

Spectra 3

Thomas A. Kubic and Assoc.
Ethyl cellulose

Spectra 4

Thomas A. Kubic and Assoc.
Cellulose triacetate

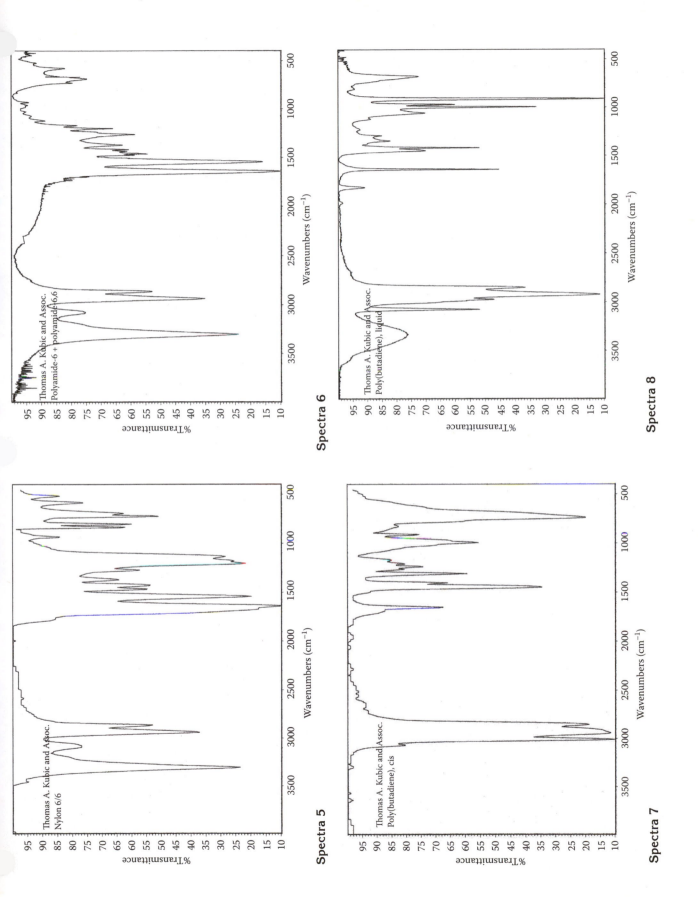

Thomas A. Kubic and Assoc.
Nylon 6/6

Spectra 5

Thomas A. Kubic and Assoc.
Polyamide-6 + polyamide 6,6

Spectra 6

Thomas A. Kubic and Assoc.
Poly(butadiene), cis

Spectra 7

Thomas A. Kubic and Assoc.
Poly(butadiene), liquid

Spectra 8

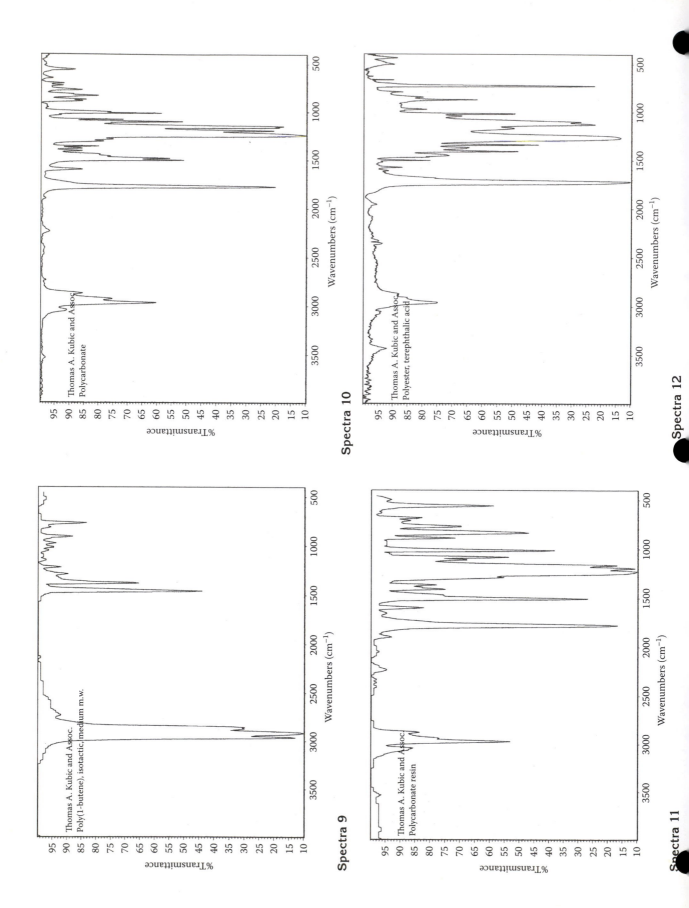

Thomas A. Kubic and Assoc.
Poly(1-butene), isotactic, medium m.w.

Spectra 9

Thomas A. Kubic and Assoc.
Polycarbonate

Spectra 10

Thomas A. Kubic and Assoc.
Polycarbonate resin

Spectra 11

Thomas A. Kubic and Assoc.
Polyester, terephthalic acid

Spectra 12

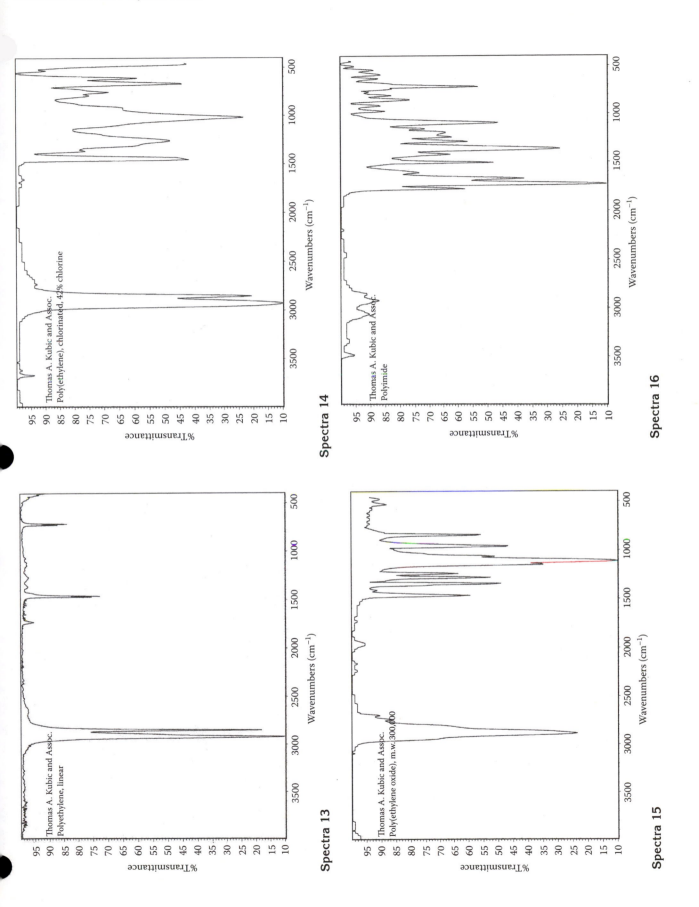

Spectra 13

Thomas A. Kubic and Assoc.
Polyethylene, linear

Spectra 14

Thomas A. Kubic and Assoc.
Poly(ethylene), chlorinated, 42% chlorine

Spectra 15

Thomas A. Kubic and Assoc.
Poly(ethylene oxide), m.w. 300,000

Spectra 16

Thomas A. Kubic and Assoc.
Polyimide

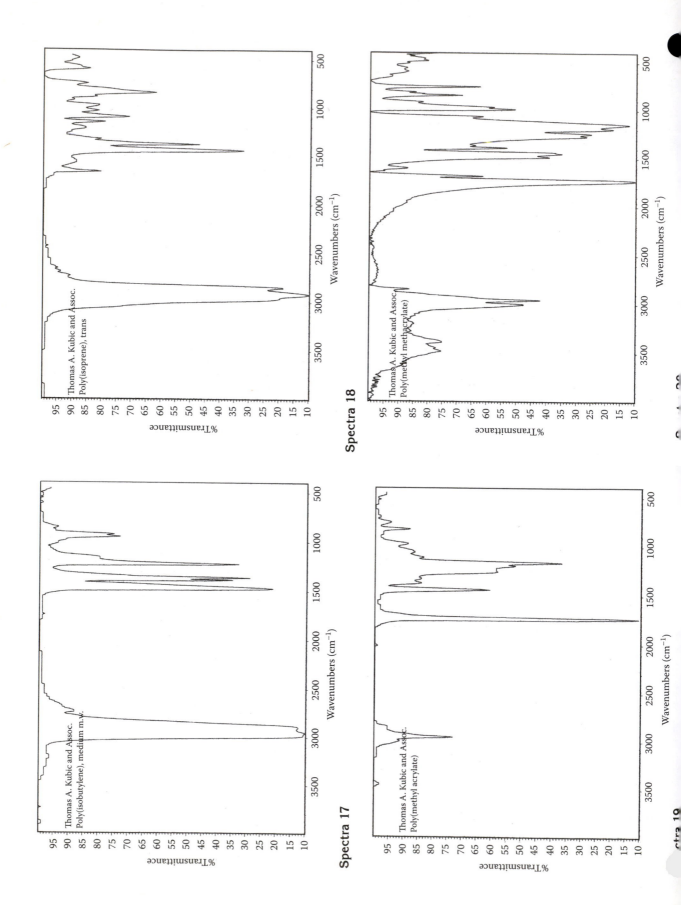

Thomas A. Kubic and Assoc.
Poly(isoprene), trans

Thomas A. Kubic and Assoc.
Poly(methyl methacrylate)

Spectra 18

Thomas A. Kubic and Assoc.
Poly(isobutylene), medium m.w.

Spectra 17

Thomas A. Kubic and Assoc.
Poly(methyl acrylate)

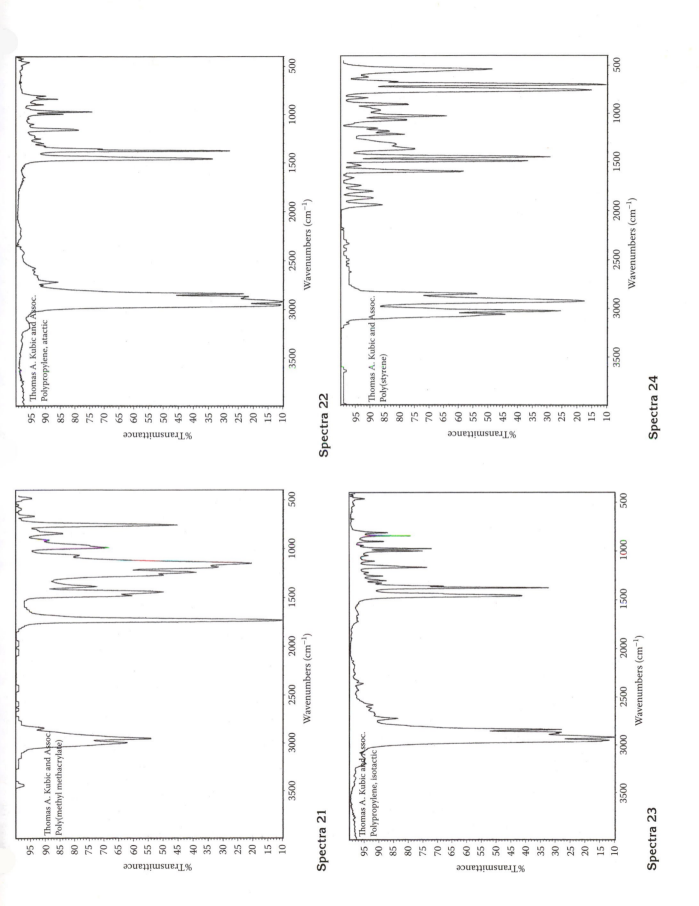

Spectra 21

Thomas A. Kubic and Assoc.
Poly(methyl methacrylate)

Spectra 22

Thomas A. Kubic and Assoc.
Polypropylene, atactic

Spectra 23

Thomas A. Kubic and Assoc.
Polypropylene, isotactic

Spectra 24

Thomas A. Kubic and Assoc.
Poly(styrene)

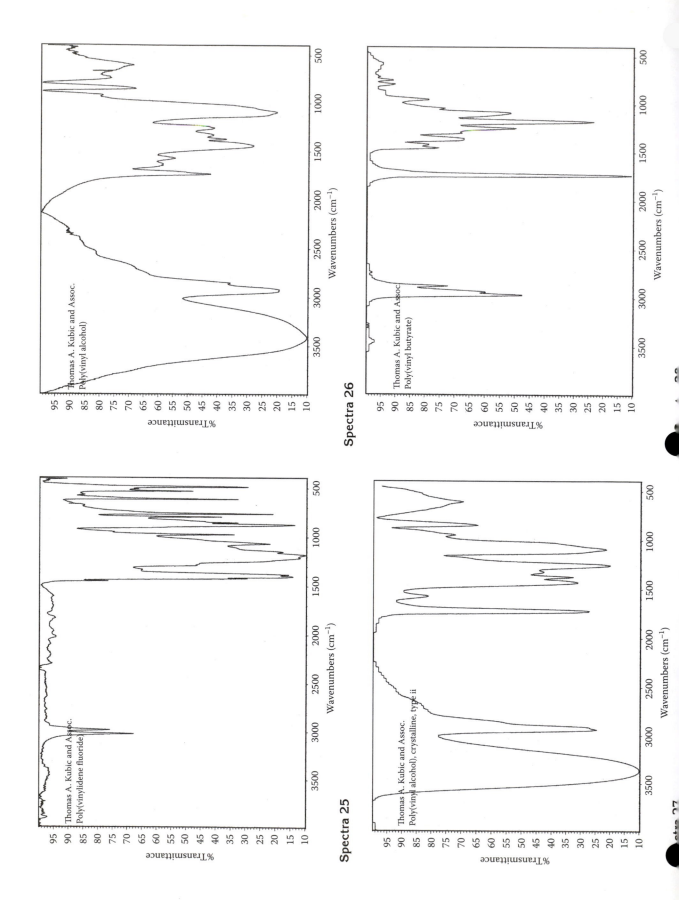

Spectra 25

Thomas A. Kubic and Assoc.
Poly(vinylidene fluoride)

Spectra 26

Thomas A. Kubic and Assoc.
Poly(vinyl alcohol)

Thomas A. Kubic and Assoc.
Poly(vinyl butyrate)

Thomas A. Kubic and Assoc.
Poly(vinyl alcohol), crystalline, type ii

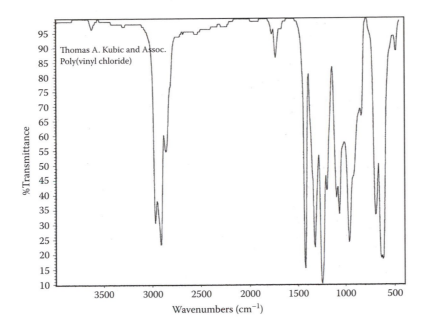

Thomas A. Kubic and Assoc.
Poly(vinyl chloride)

%Transmittance

Wavenumbers (cm^{-1})

Spectra 29

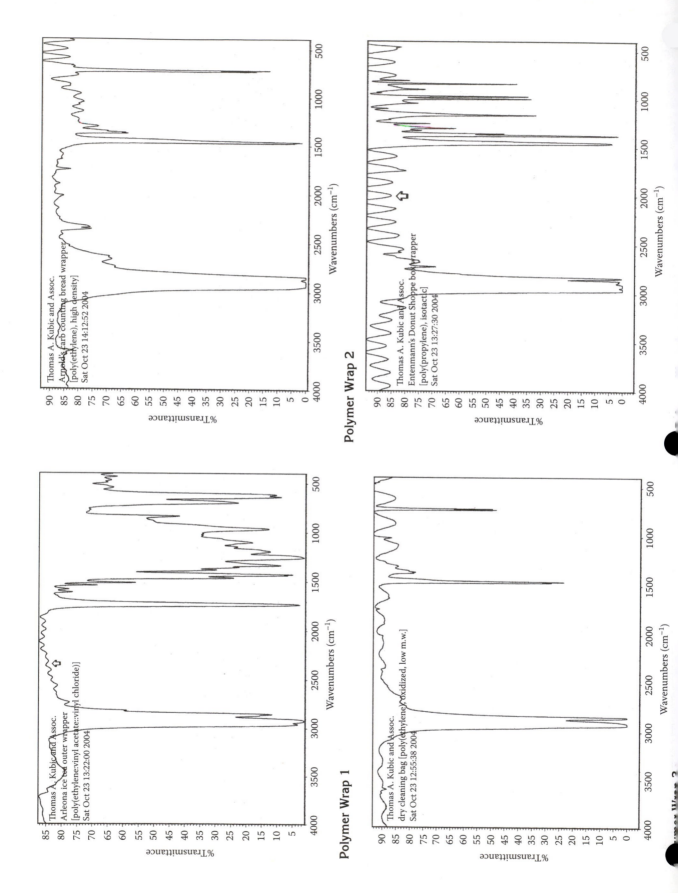

Thomas A. Kubic and Assoc.
Arnold's [arb counting bread wrapper
[poly(ethylene), high density]
Sat Oct 23 14:12:52 2004

Thomas A. Kubic and Assoc.
Entenmann's Donut Shoppe bow wrapper
[poly(propylene), isotactic]
Sat Oct 23 13:27:30 2004

Polymer Wrap 2

Thomas A. Kubic and Assoc.
Arleona ice bel outer wrapper
[poly(ethylene:vinyl acetate:vinyl chloride)]
Sat Oct 23 13:22:00 2004

Polymer Wrap 1

Thomas A. Kubic and Assoc.
dry cleaning bag [poly(ethylene), oxidized, low m.w.]
Sat Oct 23 12:55:38 2004

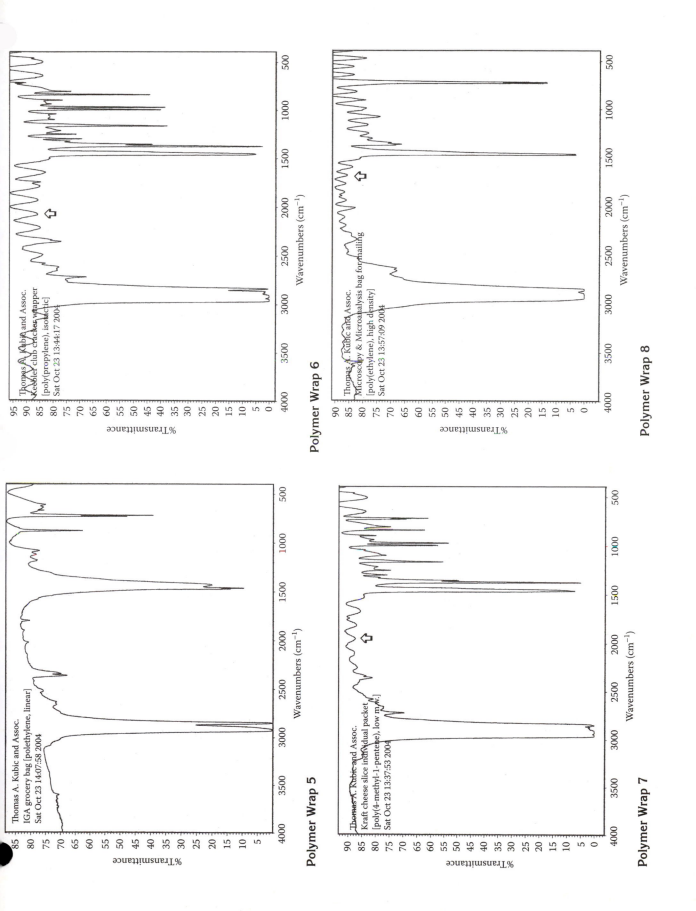

Thomas A. Kubic and Assoc.
Keebler club cracker wrapper
[poly(propylene), isotactic]
Sat Oct 23 13:44:17 2004

%Transmittance

Wavenumbers (cm^{-1})

Polymer Wrap 6

Thomas A. Kubic and Assoc.
Microscopy & Microanalysis bag for mailing
[poly(ethylene), high density]
Sat Oct 23 13:57:09 2004

%Transmittance

Wavenumbers (cm^{-1})

Polymer Wrap 8

Thomas A. Kubic and Assoc.
IGA grocery bag [polethylene, linear]
Sat Oct 23 14:07:58 2004

%Transmittance

Wavenumbers (cm^{-1})

Polymer Wrap 5

Thomas A. Kubic and Assoc.
Kraft cheese slice individual packet
[poly(4-methyl-1-pentene, low mw]
Sat Oct 23 13:37:53 2004

%Transmittance

Wavenumbers (cm^{-1})

Polymer Wrap 7

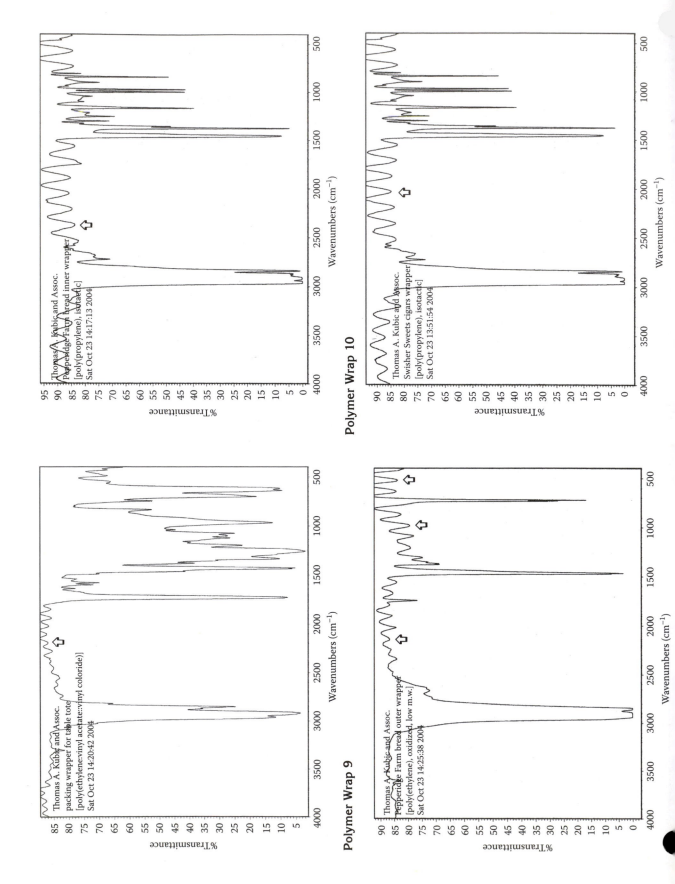

Polymer Wrap 9

Polymer Wrap 10

Thomas A. Kubic and Assoc.
Pepperidge Farm bread inner wrapper
[poly(propylene), isotactic]
Sat Oct 23 14:17:13 2004

Thomas A. Kubic and Assoc.
Swisher Sweets cigars wrapper
[poly(propylene), isotactic]
Sat Oct 23 13:51:54 2004

Thomas A. Kubic and Assoc.
packing wrapper for table tote
[poly(ethylenevinyl acetate::vinyl coloride)]
Sat Oct 23 14:20:42 2004

Thomas A. Kubic and Assoc.
Pepperidge Farm bread outer wrapper
[poly(ethylene), oxidized, low m.w.]
Sat Oct 23 14:25:38 2004

Spectra Key

1. Cellophane
2. Cellulose triacetate
3. Ethyl cellulose
4. Nylon 6
5. Nylon 6/6
6. Polyamide-6 + polyamide-6,6
7. Poly(butadiene), cis
8. Poly(butadiene), liquid
9. Poly(1-butene), isotactic, medium m.w.
10. Polycarbonate
11. Polycarbonate resin
12. Polyester, terephthalic acid
13. Polyethylene, linear
14. Poly(ethylene), chlorinated, 42% chlorine
15. Poly(ethylene oxide), m.w. 300,000
16. Polyimide
17. Poly(isobutylene), medium m.w.
18. Poly(isoprene), trans
19. Poly(methyl acrylate)
20. Poly(methyl methacrylate)
21. Poly(methyl methacrylate)
22. Polypropylene, atactic
23. Polypropylene, isotactic
24. Poly(styrene)
25. Poly(vinylidene fluoride)
26. Poly(vinyl alcohol)
27. Poly(vinyl alcohol), crystalline, type ii
28. Poly(vinyl butyrate)
29. Poly(vinyl chloride)

Polymer Wraps

1. Arleona ice tea outer wrapper [poly(ethylene:vinyl acetate::vinyl chloride)]
2. Arnold's Carb Counting bread wrapper [poly(ethylene), high density]
3. Dry cleaning bag [poly(ethylene), oxidized, low m.w.]
4. Entenmann's Donut Shoppe box wrapper [poly(propylene), isotactic]
5. IGA grocery bag [polyethylene, linear]
6. Keebler Club Cracker wrapper [poly(propylene), isotactic]
7. Kraft cheese slice individual packet [poly(4-methyl-1-pentene), low m.w.]
8. Microscopy & Microanalysis bag for mailing [poly(ethylene), high density]
9. Packing wrapper for Table Tote [poly(ethylene):vinyl acetate::vinyl chloride)]
10. Pepperidge Farm bread inner wrapper [poly(propylene), isotactic]
11. Pepperidge Farm bread outer wrapper [poly(ethylene), oxidized, low m.w.]
12. Swisher Sweets cigar wrapper [poly(propylene), isotactic]

Bibliography

Abbott, J.R., *Footwear Evidence*, Charles C Thomas, Springfield, IL, 1964.

Adoryan, A.S., and Kolenosky, G.B., *A Manual for the Identification of Hairs of Selected Ontario Mammals,* Research Report (Wildlife) no. 90, Department of Lands and Forests, Ottawa, Ontario, Canada, 1969.

American Association of Textile Chemists and Colorists, *Technical Manual*, vol. 53, American Association of Textile Chemists and Colorists, Research Triangle Park, NC, 1977.

Appleyard, H.M., *Guide to the Identification of Animal Fibres*, 2nd ed., Wira House, Leeds, U.K., 1978.

Banner, H., and Conan, B.J., *The Identification of Mammalian Hair*, Inkata Press, Melbourne, Australia, 1974.

Bass, W.M., *Human Osteology—A Laboratory and Field Manual,* 3rd ed., Missouri Archaeological Society, Columbia, MO, 1995.

Bisbing, R.E., The forensic identification and association of human hair, in *Forensic Science Handbook*, R. Saferstein, ed., Prentice-Hall, Englewood Cliffs, NJ, 1982.

Bodziak, W.J., *Footwear Impression Evidence*, 2nd ed., CRC Press, Boca Raton, FL, 2000.

Brown, F.M., The microscopy of mammalian hair for anthropologists, *Proc. Am. Philos. Soc.*, 85, 250, 1942.

Byers, S.N., *Introduction to Forensic Anthropology—A Textbook*, Allyn & Bacon, Boston, 2002.

Cassidy, M.J., *Footwear Identification*, Public Relations Branch of the Royal Canadian Police, Ottawa, Ontario, Canada, 1980.

Catling, D.M., and Grayson, J.E., *Identification of Vegetable Fibres*, Chapman & Hall, London, 1982.

Coyote, W.A., ed., *Papermaking Fibers: A Photomicrographic Atlas*, Syracuse University Press, Syracuse, New York, 1980.

Crown, D., *The Forensic Examination of Paint and Pigments*, Charles C Thomas, Springfield, IL, 1968.

Davis, J.E., *An Introduction to Tool Marks, Firearms and the Striagraph*, Charles C Thomas, Springfield, IL, 1958.

De Forest, P.R., Gaensslen, R.E., and Lee, H.C., *Forensic Science—An Introduction to Criminalistics*, McGraw-Hill, New York, 1983.

Federal Bureau of Investigation, *The Science of Fingerprints: Classification and Uses,* U.S. Government Printing Office, Washington, DC, 1998.

Field, R.M., *An Outline of the Principles of Geology*, 3rd ed., Barnes & Noble, New York, 1938.

Frei-Sulzer, M., Coloured fibres in criminal investigations with special reference to natural fibres, in *Methods of Forensic Science*, vol. IV, A.S. Curry, ed., Interscience, New York, 1965, pp. 141–175.

Fulton, C.C., *Modern Microcrystal Tests for Drugs,* John Wiley & Sons, New York, 1969.

Given, B.W., Nehrich, R.B., and Shields, J.C., *Tire Tracks and Tread Marks*, Gulf, Houston, TX, 1977.

Glaister, J., *A Study of Hairs and Wools Belonging to the Mammalian Group of Animals, Including a Special Study of Human Hair, Considered from Medico-Legal Aspects,* MISR Press, Cairo, Egypt, 1931.

Graves, W.J., Mineralogical soil classification technique for the forensic scientist, *J. Forensic Sci.*, 24, 331–337, 1979.

Gross, H., *Criminal Investigation,* adapted from *System Der Kriminalistik*, by J.C. Adams, Sweet and Maxwell Ltd., London, 1924, pp. 131–138.

Hamm, E.D., The individuality of class characteristics in Converse All-Star footwear, *J. Forensic Ident.*, 39, 277–292, 1989.

Hamm, E.D., Track identification: a historical overview, *J. Forensic Ident.*, 39, 333–338, 1989.

Hatcher, J.S., Jury, J.R., and Weller, J.A.C., *Firearms Investigation Identification and Evidence*, 2nd ed., Stackpole Co., Harrisburg, PA, 1977.

Hausman, L.H., Structural characteristics of the hair of mammals, *Am. Nat.,* 54, 496, 1920.

Hicks, J.W., *Microscopy of Hair*, U.S. Government Printing Office, Washington, DC, 1977.

Hilton, O., *Scientific Examination of Questioned Documents*, Elsevier, New York, 1982.

James. S.H., and Nordby, J.J., eds., *Forensic Science—An Introduction to Scientific and Investigative Techniques*, CRC Press, Boca Raton, FL, 2003.

Kirk, P.L., *Crime Investigation*, Interscience Publishers, New York, 1953.

Kirk. P.L., *Density and Refractive Index*, Charles C Thomas, Springfield, IL, 1951.

Krogman, W.M., *The Human Skeleton in Forensic Medicine*, Charles C Thomas, Springfield, IL, 1962.

Kubic, T., and Petraco, N., Microanalysis and examination of trace evidence, in *Forensic Science—An Introduction to Scientific and Investigative Techniques*, S.H. James and J.J. Nordby, eds., CRC Press, Boca Raton, FL, 2003.

Lee, H.C., and Gaensslen, R.E., eds., *Advances in Fingerprint Technology*, Elsevier, New York, 1991.

Lesko, J., Torpey, F., Kelly, J.P., and Behr, F.O., *City of New York Police Department Crime Scene Technicians Manual*, New York City Police Department, New York, 1977.

Locard, E., The analysis of dust traces, part I, *Am. J. Police Sci.*, 1, 276–298, 1930.

Locard, E., The analysis of dust traces, part II, *Am. J. Police Sci.*, 1, 405–406, 1930.

Locard, E., The analysis of dust traces, part III, *Am. J. Police Sci.*, 1, 496–514, 1930.

Longhetti, A., and Roche, G.W., Microscopic identification of man-made fibers from the criminalistics point of view, *J. Forensic Sci.* 3, 303–229, 1958.

Man-Made Fiber Producers Association, *Man-Made Fibers Fact Book*, Man-Made Fiber Producers Association, Washington, DC, 1978.

McCrone, W.C., Particle analysis in the crime laboratory, in *The Particle Atlas*, vol. 5, W.C. McCrone, J.G. Delly, and S.J. Pelanik, eds., Ann Arbor Science Publishers, Ann Arbor, MI, 1979, p. 1383.

McCrone, W.C. and Delly, J.D., *The Particle Atlas*, 2nd ed., Ann Arbor Science Publishers, Ann Arbor, MI, 1973.

Miller, E.T., Forensic glass comparisons, in *Forensic Science Handbook*, R. Saferstein, ed., Prentice-Hall, Englewood Cliffs, NJ, 1982, p. 154.

Moncrief, R.W., *Man-Made Fibers*, 6th ed., Newnes-Butterworths, London, 1975.

Moore, T.D., Spence, L.E., Dugnolle, C.E., and Hepworth, W.G., *Identification of the Dorsal Guard Hairs of Som Mammals of Wyoming*, Bulletin no. 14, Department of Fish and Game. Cheyenne, WY, 1974.

Murray, R.C., and Tedrow, J.C.F., *Forensic Geology*, Rutgers University Press, New Brunswick, NJ, 1975.

National Police Agency (Tokyo), An electrostatic method for lifting footprints, *Int. Criminal Police Rev.*, 272, 287–292, 1973.

Ojena, S.M., and De Forest P., Precise refractive index determinations by the immersion method, using phase contrast microscopy and the Mettler hotstage, *J. Forensic Sci. Soc.*, 12, 315–329, 1972.

Olsen, R.D., Sr., *Scott's Fingerprint Mechanics*, Charles C Thomas, Springfield, IL, 1978.

Palenik, S. Microscopy and microchemistry of physical evidence, in *Forensic Science Handbook,* vol. II, R. Saferstein ed., Prentice-Hall, Englewood Cliffs, NJ, 1988, pp. 165–167.

Palenik, S.J., Microscopical examination of fibers, in *Forensic Examination of Fibers*, J. Robertson and M. Grieve, eds., Taylor and Francis, Philadelphia, 1999.

Parham, R.A., and Gray, R.L., *The Practical Identification of Wood Pulp Fibers*, Tappi Press, Atlanta, GA, 1982.

Petraco, N., A guide to the rapid screening, identification, and comparison of synthetic fibers in dust samples, *J. Forensic Sci.*, 32, 768–777, 1987.

Petraco, N., A microscopical method to aid in the identification of animal hair, *Microscope*, 35, 83–92, 1987.

Petraco, N., A modified technique for the cross-sectioning of hairs and fibers, *J. Police Sci. Admin.*, 9, 448–450, 1981.

Petraco, N., The replication of hair cuticle scale patterns in Meltmount®, *Microscope*, 34, 341–345, 1986.

Petraco, N., Trace evidence—the invisible witness, *J. Forensic Sci.*, 31, 321–328, 1986.

Petraco, N., and De Forest, P.R., A guide to the analysis of forensic dust specimens, in *Forensic Science Handbook*, vol. III, R. Saferstein, ed., Prentice Hall, Englewood Cliffs, NJ, 1993.

Petraco, N., and Kubic, T., *Color Atlas and Manual of Microscopy for Criminalists, Chemists and Conservators*, CRC Press, Boca Raton, FL, 2003.

Petraco, N., Resua, R., and Harris, H.H., A rapid method for the preparation of transparent footwear test prints, *J. Forensic Sci*, 27, 935–937, 1982.

Pickering, R.B., and Bachman, D.C., *The Use of Forensic Anthropology*, CRC Press, Boca Raton, FL, 1997.

Robertson, J., ed., *Forensic Examination of Fibres*, Ellis Horwood, Chichester, U.K., 1992.

Robertson, J., The forensic examination of fibres: protocols and approaches—an overview, in *Forensic Examination of Fibres*, J. Robertson, ed., Ellis Horwood, Chichester, U.K., 1992.

Robertson, J., ed., *Forensic Examination of Hair,* Taylor and Francis, London, 1999.

Robertson, J., and Grieve, M., eds., *Forensic Examination of Fibers*, Taylor and Francis, Philadelphia, 1999.

Rouessac, F., and Rouessac, A., *Chemical Analysis: Modern Instrumentation Methods and Techniques*, John Wiley & Sons, New York, 2000.

Russ, J.C., *Forensic Uses of Digital Imaging*, CRC Press, Boca Raton, FL, 2001.

Sansone, S.J., *Police Photography,* Anderson, Cincinnati, OH, 1977.

Sato, H., Yoshino, M., and Seta, S., Macroscopical and microscopical studies of mammalian hairs with special reference to the morphological differences, *Rep. Natl. Res. Inst. Police Sci.*, 33, 1–16, 1980.

Scott, C.C., *Photographic Evidence*, 2nd ed., vol. 1, West Publishing, St. Paul, MN, 1969.

Shaffer, S.A., A protocol for the examination of hair evidence, *Microscope*, 30, 151–161, 1982.

Skoog, D.A., Holler, F.J., and Nieman, T.A., *Principals of Instrumental Analysis*, 5th ed., Harcourt Brace College, Philadelphia, 1998.

Söderman, H., and Fontell, E., *Handbok I. Kriminalteknik,* Stockholm, 1930, pp. 534–552.

Smith, E.J., *Principles of Forensic Handwriting Identification and Testimony*, Charles C Thomas, Springfield, IL, 1984.

Smith, S., and Glaister, J., *Recent Advances in Forensic Medicine*, 2nd ed., Blakiston's Son & Co., Philadelphia, 1939, pp. 118–124.

Svensson, A., Wendel, O., and Fisher, B.A.J., *Techniques of Crime Scene Investigation*, 3rd ed., American Elsevier, New York, 1981.

Technical Working Group on Materials Analysis, Fiber Subgroup, *Forensic Fiber Examination Guidelines*, Technical Working Group on Materials Analysis, Washington, DC, 1998.

Textile Institute, *Identification of Textile Materials*, Manchester, Textile Institute, 1970.

U.S. Department of Commerce, Reference collection of synthetic fibers, National Bureau of Standards, Washington, DC, January 1984.

Wildman, A.B., *Microscopy of Animal Textile Fibres*, Wira House, Leeds, U.K., 1954.

Zeno, G., Use of computers in forensic science, in *Forensic Science—An Introduction to Scientific and Investigative Techniques*, S.H. James and J.J. Nordby, eds., CRC Press, Boca Raton, FL, 2003.